ISAAC ASIMOV is undoubtedly America's foremost writer of science for the layman. An Associate Professor of Biochemistry at the Boston University School of Medicine, he is the author of such standards of science reportage as THE GENETIC CODE, THE UNIVERSE, FACT AND FANCY, LIFE AND ENERGY, and the three-volume UNDERSTANDING PHYSICS, in addition to many science fiction works which are considered classics of the genre. Born in Russia, Isaac Asimov came to this country with his parents at the áge of three, and grew up in Brooklyn. In 1948 he received his Ph.D. in Chemistry at Columbia and joined the faculty at Boston University. In its review of this book, *Choice* called Asimov "first-rate ... a master explainer; accurate, profound, and at the same time exciting."

Other Avon books by
Isaac Asimov

THE SOLAR
SYSTEM AND BACK

Isaac Asimov

 A DISCUS BOOK/PUBLISHED BY AVON BOOKS

All essays in this volume are reprinted from *The Magazine of Fantasy and Science Fiction*. Individual essays appeared in the following issues:

AVON BOOKS
A division of
The Hearst Corporation
959 Eighth Avenue
New York, New York 10019

Copyright © 1970 by Doubleday & Company, Inc.
Published by arrangement with Doubleday & Company, Inc.
Library of Congress Catalog Card Number: 78-89121.
ISBN: 0-380-01444-0

First Discus Printing May, 1972
Fourth Printing

DISCUS BOOKS TRADEMARK REG. U.S. PAT. OFF. AND
FOREIGN COUNTRIES, REGISTERED TRADEMARK—
MARCA REGISTRADA, HECHO EN CHICAGO, U.S.A.

Printed in the U.S.A.

To—Howard Gotlieb,
curator extraordinaire

CONTENTS

II — AND BACK

Introduction

Writing the essays that fill this book and its predecessors, and that, prior to appearing in book form, are to be found in each issue of *The Magazine of Fantasy and Science Fiction,* has certain aspects of a game to me.

The game is this: I do my best to treat a subject accurately and concisely and the readers do their best to catch me out in errors. The readers often score points, though rarely disastrous ones, and I make the necessary corrections in public.

Occasionally, there *is* a disaster that precludes republication of a particular essay altogether, but I'm happy to say that hasn't happened more than twice in over ten years of essay writing, and now I want to take back the adverse decision in one of those cases. Where can I do this better and more happily than in the introduction to one of these collections?

The essay in question appeared in early 1959. Here it is:

NOTHING

The word "vacuum" comes from the Latin *vacuus*, meaning "empty." (We run close to the original in nonscientific lingo, too, as when we talk of a "vacuous stare.") Consequently, a vacuum is empty space, or space that contains nothing.

This is fine, in theory, and we talk easily of vacuums when all we mean is a volume of space with less matter in it than we are accustomed to. The question is, though: Does a true, ideal vacuum exist? Or, to put it another way: Is there such a thing as nothing?

In between atoms of a gas there is no matter in the ordinary sense (there may be stray electrons or neutrinos) so we can speak of an "interatomic vacuum." However, if we raise ourselves above the atomic level and consider a reasonable volume of the universe, say a cubic centimeter, then the question of a vacuum grows more interesting.

For instance, at sea-level pressure and 20° C. (68° F.), air has a density of 0.0012 gram per cubic centimeter. This means that every cubic centimeter contains 2.5×10^{19} molecules of oxygen and nitrogen. Almost all the mass of the molecules is in the protons and neutrons (together called nucleons) in the nuclei of the atoms making them up. The oxygen molecule contains 32 nucleons and the nitrogen molecule has 28. There are four nitrogen molecules for every oxygen molecule in the atmosphere, so, altogether, ordinary air contains 7.25×10^{20} nucleons per cubic centimeter.

In the laboratory, it is possible to prepare a volume of space from which nearly all the air has been evacuated. This is *called* a vacuum, and in very good man-made vacuums the amount of air left is only about a ten-billionth the original quantity.

That's not bad, you understand, but even such a man-made vacuum retains about 7.25×10^{10} nucleons per cubic centimeter. That's nearly a hundred billion nucleons in every cubic centimeter and that sounds as though there's still a bit of crowding going on.

Of course, the science-fiction reader knows a trick worth several times this. There is always the "vacuum of outer space."

After all, as we move up away from the Earth, the

atmosphere thins out, and at a height of 200 miles or so, it becomes a vacuum that is at least as good as any we can make in the laboratory, and at a height still greater, the vacuum is still better (or to use the appropriate jargon, it is still "harder").

But even though interplanetary space is a fine vacuum compared to the miserable specimens we can manage, it is still far from nothing. There is the debris left over from the dust and gas out of which the solar system was formed. In fact, even in interstellar space, the vast reaches between the stars, there is still debris left over from the original dust and gas used in the formation of the stars. The interstellar matter is thick enough to raise obscuring black clouds out in the galactic arms, where our Sun is located and where most of the dust exists (as compared with the comparatively dustless galactic center). The case is similar for other galaxies.

The average density of matter in interstellar space is 10^{-21} gram per cubic centimeter according to some (necessarily rough) estimates I have seen. This would amount to only 1000 nucleons per cubic centimeter. Interstellar space is a vacuum that is nearly a hundred million times as hard as any we can manage but it obviously isn't nothing.

Still, we have one more trick to play. What about the stupendous distances between the galaxies, distances that dwarf the already tremendous stretches between the stars within a galaxy. Surely, intergalactic space ought to be emptier than interstellar space—and it is.

But yet, even intergalactic space is not quite nothing. Astronomers still detect some matter there, enough to leave its mark on light reaching us from distant galaxies.

But if intergalactic space is not nothing, how near nothing is it? The lowest figure I've seen for the density of the matter in intergalactic space is (and again I warn you, it's only a rough approximation) 10^{-24} gram per cubic centimeter. This amounts to just about 1 nucleon per cubic centimeter.

This is as close to nothing as anything in the universe ever gets (or, perhaps, ever can get). It's *so* close to nothing that it would seem at first glance that we can just call it nothing and forget about it.

But can we forget about it? Is it so close to nothing, it doesn't matter? Let's see.

First, space is large and we needn't confine ourselves to pinches of it. Suppose we take a volume the size of the Earth and imagine it filled with intergalactic matter. The Earth takes up a large number of cubic centimeters, to be sure—1.1×10^{27}, to be exact. There would be 1 nucleon for each of those and the total mass of these would come to 1800 grams (4 pounds).

The Sun, with a volume that is 1,300,000 times that of the Earth, would, if composed of intergalactic material, contain a mass of 2.3×10^9 grams or 2600 *tons*.

However, it is unfair to try to get an idea of intergalactic matter by using such small units of volume. The Earth, and even the Sun, are submicroscopic dots compared to the universe, and their volumes are beneath contempt.

We have all space to consider. In measuring distances outside out solar system, the smallest useful unit is the light-year. Surely, then, in measuring spatial volume, we ought to use cubic light-years as the minimum unit. (A cubic light-year is, of course, a cube which is 1 light-year, or 9.5 trillion kilometers, on each side.)

A cubic light-year contains 8.5×10^{53} cubic centimeters. One nucleon for each of that vast number of cubic centimeters comes to a mass of 1.4×10^{30} grams or 1.5×10^{24} (one and a half trillion trillion) *tons*.

Now we have something! This hardest of all hard vacuums, this nothingest of all nothings, still piles up matter in the trillions of trillions of tons when a volume of a cubic light-year is involved. And, after all, a cubic light-year, large as it is from a merely human standpoint, is really an insignificant fraction of the volume of the universe.

To show you what I mean, consider that the 200-inch telescope can penetrate over a billion light-years in every direction and, as far as we can see or photograph, galaxies stretch out. There is no way of telling yet how much further the universe reaches but let's imagine a sphere with ourselves at the center and a radius of a billion light-years. We'll content ourselves with this (no doubt, tiny) fraction of all and call it the "observable universe."

With a radius of a billion (10^9) light-years, the volume of the observable universe can be calculated easily enough. It turns out to be 4×10^{27} (four thousand trillion trillion) cubic light-years. You see, I was right in saying that a single cubic light-year is really a mere yawn-worthy speck.

(Not all the universe is intergalactic space, of course; some of it consists of the galaxies themselves. However, the galaxies make up only a tenth of a per cent of all space, so we can ignore them at this point.)

At the rate of 1 nucleon per cubic centimeter, the amount of matter in the intergalactic space of the observable universe comes to 5.6×10^{57} grams.

This sounds like a lot, and 5,600,000,000,000,000,-000,000,000,000,000,000,000,000,000,000,000,000 grams *is* a lot. How much is it, though, on a universal scale? For instance: How much is it in comparison to the mass of a star? Our own Sun, an average star, has a mass of 2×10^{33} grams, so the intergalactic matter weighs much more than a star; much, much more. In fact, the amount of intergalactic matter in a mere couple of thousand cubic light-years equals the mass of the Sun.

However, the Sun is only one of many. Our galaxy contains about a hundred billion (10^{11}) stars. The total mass of all the stars in the galaxy is therefore about 2×10^{44} grams, assuming the Sun's mass to be the average for stars, which it probably is. As you see, the inter-galactic matter weighs more than the stars in an entire galaxy; much more.

But our galaxy is also only one of many. In the observable universe, it is estimated that there are about a hundred billion (10^{11}) galaxies. The total mass of all the stars in the observable universe is therefore 2×10^{55} grams.

So, as it turns out, the intergalactic matter is *still* more massive, more massive than all the stars in all the galaxies

—280 times as massive, if all the round approximations I have used are considered accurate.

In fact, a godlike creature from outside our universe, looking over the entire business with a casual eye, would be justified in describing it as nothing more than a hard vacuum. He would be just as right to ignore the occasional dots of non-vacuum, as we are in ignoring dust particles when we describe our atmosphere as a gas.

Is the vacuum of intergalactic space nothing?

Heck, no! If we consider only quantity—it is practically everything.

*   *   *

After this appeared, I received a letter from a young astronomer friend of mine who told me that my estimate of the density of the intergalactic gas was far too high on the basis of recent data and that the mass of dispersed intergalactic matter was probably not more than 2 per cent that of the galaxies. I sighed, and retired the article.

And then, in early 1968, nine years after the article was written, new evidence turned up. By measuring X-ray fluxes in outer space from rockets above the atmosphere, scientists at White Sands Proving Ground, New Mexico, decided there could well be a surprising quantity of hot hydrogen in intergalactic space.

In fact, they say, "there may be one hundred times as much matter dispersed as a gas in the vast reaches of space between galaxies as there is in all the mass of all the galaxies combined."

This is not yet as high as the figure I arrived at by guess and by figuring, and the initial X-ray evidence may turn out to be in error or to be misleading. Still, it is enough to make me bring my old article out of retirement.

And to make me feel very good, too.

I
THE
SOLAR
SYSTEM

A
THE INNER SYSTEM

1 The Seventh Planet

Every once in a while, as my Gentle Readers know, I pull a blooper in one of my essays. In that case, a number of Readers write Gently to say: "Doesn't five plus four come to nine? You said eight!" Then I correct myself with an embarrassed giggle.*

To err is human, and to correct gently is humane.

But there also comes a time when it is necessary for me to correct an article because scientists have discovered *they* had made a mistake. Then, unaccountably, I am furious. How *dare* they make a mistake!

A case happened a couple of years ago and I've been brooding about it ever since. In my essay "Round and Round and—" (which appeared in *Of Time and Space and Other Things,* Doubleday, 1965), I made the casual statement: "Both Mercury and Venus turn one face eternally to the Sun. . . ."

That was exactly right as far as I, or anyone else in the world, knew when that article was written (in mid-1963 actually), but it is no longer right. Astronomers have

* See, however, the Introduction, in case you skipped it.

changed their minds, and the back of my hand to them. They not only outdated an article, but also two novels and one novelette that I had written with painstaking attention to scientific accuracy. Have they no heart?

But it is an ill wind indeed, out of which I cannot make an essay, and it is time now for me to consider the new situation in some detail. As is usual for me, I will begin at the beginning and deal with the seven planets known to the ancients.

The Greeks considered any body that moved, relative to the stars, to be a planet, and therefore included the Sun and Moon among their number. The remaining five, which are bright starlike objects more or less easily visible to the naked eye, are Mars, Jupiter, Saturn, Venus, and Mercury.

Of these five, three are farther from the Sun than the Earth is. This means that they move in great swings about Sun and Earth alike. They can each of them be so placed that the Earth is directly between them and the Sun. When this is so, the planet is at the zenith when the Sun is at nadir on the other side of the Earth. This means that the planet is at zenith at midnight. Obviously, any bright starlike object which is high in the sky at midnight is easy to observe.

The case is quite different for Venus and Mercury, which are closer to the Sun than the Earth is. They can *never* be in such a position that the earth is directly between them and the Sun, because the Earth would then have to be closer to the Sun than they are and that is not so. This means that Venus and Mercury can *never* be seen in the night sky at the zenith.

It can happen, on the other hand, that either Venus or Mercury is more or less directly between the Earth and the Sun. This is called "inferior conjunction" and each planet is then closer to the Earth than it ever is at other times. Venus will be as close to us as 25,000,000 miles and Mercury as close as 49,000,000. The only trouble is that in that case, in order to see either planet we must look in the direction of the Sun and everything is lost in the glare.

Of course, if either planet is *exactly* between ourselves and the Sun, it will show up against the Sun's disk. But then it is the side away from us (and toward the Sun) that is lit by sunlight, and all we see is the black disk of the night side against the brightness of the Sun.

But let's concentrate on Venus to begin with, and follow its travels about the Sun. We can start with it at inferior conjunction between ourselves and the Sun.

Steadily, as it moves in its orbit, Venus pulls away from the Sun. (We move also, and in the same direction, but Venus moves more quickly than Earth does, being closer to the Sun's pull.) As it moves away, we can see around the night side a trifling bit and the edge of the sunlit side (as seen through a telescope) looks like a thin crescent.

The farther it gets away from the Sun, the more of the sunlit side we can see and the thicker the crescent. Finally, the imaginary line connecting Venus to the Sun comes to be at right angles to the imaginary line connecting Venus to the Earth. We then see Venus "in profile" so to speak. Half of the face we see is sunlit, the other half is dark. The planet is in the "half-Venus" phase (see Figure 1).

Let's freeze matters at this half-Venus phase for a while. At this point in its orbit, Venus is separated from the Sun by an angular distance of 47° (as viewed from Earth). From our earthly viewing station, this is as far as Venus can possibly get from the Sun. The planet is at "maximum elongation."

Now let's go on and allow Venus to move again. As it continues in its orbit, it begins to curve away from us and back toward the Sun.

We see more and more of the sunlit side as it moves along the far side of its orbit, until just before it passes behind the Sun, the side toward us is fully lit up. We would then see "full-Venus" if we could see it at all, which we couldn't, for once again we would have to look directly at the Sun to see it. It is now at "superior conjunction."

But then it moves away from the Sun in the other

FIGURE 1 ORBIT AND PHASES OF VENUS

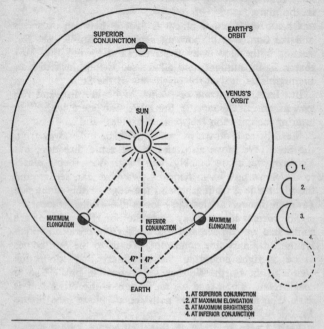

1. AT SUPERIOR CONJUNCTION
2. AT MAXIMUM ELONGATION
3. AT MAXIMUM BRIGHTNESS
4. AT INFERIOR CONJUNCTION

direction, changing from full-Venus back to half-Venus as it goes. When it is at half-Venus it is at maximum elongation again, 47°, but on the other side of the Sun. Again it moves toward the Sun, becoming a narrowing crescent until it is directly between us and the Sun again.

Let's see what a 47° maximum elongation means. The Earth makes one complete turn (360°) in 24 hours, and therefore turns through 15° in 1 hour. It turns through 47° in 3 hours and 8 minutes.

Suppose, then, that Venus is 47° east of the Sun. At sunset, Venus would still be 47° above the western horizon (halfway to zenith). In the darkening twilight it would shine out brilliantly before any but the very brightest stars makes its appearance. It is the evening star.

But three hours after sunset, the turning Earth has

caught up with it and it sets. Venus can never be seen, in the ordinary manner of speaking, higher than halfway to zenith, or for longer than three hours after sunset. (Venus is bright enough to be seen on occasion when the Sun is in the sky, but only if you know where to look and only just barely. Let's not count that, nor let us count observation by specialized instruments in broad daylight.)

If Venus, as evening star, is anywhere but at maximum elongation, it is lower in the sky and can be seen for correspondingly shorter times after sunset.

When Venus is at maximum elongation on the other side of the Sun; that is, to its west, the situation is different. It sets three hours before the Sun and isn't seen in the evening at all. Comes morning, however, and Venus rises in the east three hours before the Sun and reaches a point halfway to zenith by sunrise. Before sunrise it is shining brightly in the dawn as the glorious morning star.

It took the ancients some time to see that the evening star and the morning star were not two different objects. It was eventually borne in on observers that when the evening star was present in the sky, the morning star was absent, and vice versa. They also noted that when the evening star moved close to the Sun, there was a wait of several days and then the morning star appeared close to the other side of the Sun. The two were seen to be a single planet.

Next, let's consider the phases of Venus. In a way, the phase situation is frustrating. When Venus is near the full so that as much of it can be seen as possible, it is well on the other side of the Sun and is therefore some 150,000,000 miles away. We can only see it as a comparatively tiny object.

Then, as it approaches close to us on this side of the Sun and gets to be at only one-sixth the distance it is at maximum, we see mostly the dark side. The closer Venus gets, the less of its visible surface is sunlit. (Of course, Venus is covered by a perpetual cloud layer so that we can't see anything anyway, but it's the principle of the thing that is so exasperating.)

As a result, in considering the brightness of Venus, we must take into account two opposed effects. As it moves farther from us, it gets dimmer because of increasing distance, and brighter because of increasing sunlit area. We have to strike some compromise and it turns out that Venus is brightest when it is in a fat crescent stage, between maximum elongation and inferior conjunction. At that moment its magnitude is −4.3, or just about ten times as bright as Sirius, the brightest star. It is, in fact, the third brightest object in the heavens, surpassed only by the Sun and the Moon, and, on a clear, moonless night is said to be bright enough to cast a very dim shadow.

The mere fact that Venus's phases exist at all, however, played an important role in scientific history.

In 1543, Copernicus advanced the heliocentric theory of the solar system, suggesting that the planets, including the Earth, revolved about the Sun, instead of having all the planets, including the Sun, revolve about the Earth.

It took nearly a century for the Copernican theory to be accepted by astronomers generally. Usually, this is looked upon as a measure of the bigotry and general nastiness of Big Science, but it wasn't. The Copernican theory was a somewhat more convenient way of calculating planetary positions into the future. The mathematics was a bit easier than in the case of the Ptolemaic system. Still, because it was a mathematical shortcut, did that mean it also portrayed the solar system as it was?

To consider Copernicanism physically true as well as mathematically convenient, some observational evidence was needed, and *none existed!* In fact, since stars did not show any parallax, as they ought to have done if the Copernican theory were correct, it could be argued that there was at least one piece of observation that was *against* Copernicus. (Copernicus maintained that there were indeed parallaxes but they were too small to be measured. He was right, but the observational evidence for that was not collected until three centuries after his death.)

Without observational evidence, why should anyone believe that the firm Earth beneath his feet was flying about the Sun without his being aware of it? Had I been living in the sixteenth century, I wouldn't have believed any such cock-and-bull story, either.

But then came Galileo and his telescope. In January, 1610, he found that Jupiter had four satellites circling it. This was a kind of blow to the Ptolemaic view since it showed that there were at least four objects in the heavens that were manifestly circling some body other than Earth. It was not fatal, though, for it could easily be maintained that Jupiter and its four satellites were a single system that circled the Earth. No other point of Ptolemaic theory would be affected.

It was Venus that was the crucial planet. The only way the Ptolemaic theory could have Venus traveling about the Earth and still move in the heavens as it did (using the orbital mechanisms they insisted on using) was to arrange matters whereby Venus oscillated back and forth from one side of the Sun to the other, while always remaining *between the Sun and the Earth*. This would mean that Venus would always be at half-Venus or less; it would always be some sort of crescent, thick or thin.

On the other hand, by Copernican theory, Venus would travel completely around the Sun, and therefore, as described earlier in the article, would display all the phases from new to full, in just the manner our own Moon does.

In 1543, there were no telescopes and no one could tell whether Venus showed phases at all, let alone what kind. So when Copernican theory was born, the phases of Venus could not serve as observational evidence.

But then, in September, 1610, eight months after he had discovered the satellites of Jupiter, Galileo turned his telescope on Venus and found it *more* than half-Venus.

Galileo was enough of a scientist not to want to rush into print half-cocked. It was important that he follow Venus all through its orbit and make sure that its phases followed in order, precisely according to Copernican predictions. On the other hand, he was enough of a self-centered human being not to want to lose credit for the discovery simply because he was being cautious.

He therefore published the following Latin sentence in the little newsletter he was putting out concerning his investigations. *Haec immatura a me iam frustra leguntur, o. y.*

This means "In vain were these things gathered by me today, prematurely." This is a sort of dim hint that he was onto something he was not yet ready to disclose, but the sentence was an anagram and when the letters were rearranged it gave the true nature of his discovery. (The "o. y." at the end was needed to make the anagram come out correctly.)

Rearranged, the sentence went: *Cynthiae figuras aemulatur mater amorum.*

This means "The mother of love imitates the appearance of Cynthia." This scarcely seems clearer but Galileo was being figurative in an age in which intellectual society knew their Greek myths.

To begin with, Apollo and Artemis were born on Mt. Cynthus on the island of Delos, according to the myths, and were therefore sometimes called Cynthius and Cynthia respectively. Artemis was commonly considered to be a representation of the Moon. Consequently, Cynthia is a classical epithet for the Moon. As for the "mother of love," who could that be but Aphrodite (Venus), the mother of Eros ("love").

So, in effect, what Galileo was saying was: "Venus displays phases like those of the Moon," which is what the Copernican theory required and the Ptolemaic theory forbade.

That settled Ptolemy's hash and put Copernicus in business, except for the inevitable die-hards (mostly, but not entirely, non-astronomers).

Although Venus can be seen only for limited periods after sunset or before sunrise, those limited periods are long enough to allow Venus to be a prominent object indeed.

If a planet is considered, in Greek fashion, to be any body that moves against the background of stars, then I have no doubt that the Moon was the first planet to be discovered, for observation of the Moon on any two successive nights is enough to show that it has moved against the stars.

The Sun would have been the second planet discovered. It soon became clear that the starry pattern of the heavens shifts from night to night, and that would have been blamed on the apparent motion of the Sun against the stars. Such motion causes it to blot out a gradually shifting half of the sky.

But then the less notable planets were detected, and of these Venus surely headed the list. It is by far the brightest; and it is particularly brilliant in the dying twilight and in the glimmering beginnings of dawn when the other stars are washing out and presenting little competition.

The shift of Venus with respect to the Sun is soon obvious, and with respect to the other stars, it is only slightly less obvious. Call Venus the third planet, then, historically speaking.

Mars, Jupiter, and Saturn are all visible in the night sky at all heights, up to and including zenith. Of these, Jupiter is, on the average, the brightest, over twice as bright as the star Sirius.

However, for a short time every other year, Mars is as bright as Jupiter and, on rare occasions, even a bit brighter. What's more, it is of a distinctly reddish color that is more eye-attracting than the ordinary white of Jupiter. Furthermore, Mars moves against the background of the stars some six times as rapidly as Jupiter does. It seems to me, then, that Mars must have been the fourth planet to be discovered.

Saturn is dimmer than Jupiter and moves less than half as quickly. So Jupiter is the fifth planet and Saturn the sixth in the order of discovery.

That leaves Mercury which, I maintain, was surely the seventh and last planet to be discovered by naked eye. Why? Well, it is the nearest planet to the Sun, nearer even than Venus, which means that it shows all the orbital peculiarities of Venus, but in more pronounced fashion.

Because it is closer to the Sun than Venus is, it doesn't ever move as far from the Sun (from our Earth-based view) as Venus does. Mercury's maximum elongation under the most favorable conditions is about 27°, or less than one-third of the way to the zenith. That means it can only be seen *at most* for less than two hours after sunset or before sunrise.

In fact, there is another difficulty. All the planetary orbits are ellipses, but Venus's orbit is least elliptical while Mercury's, of those planets visible to the naked eye, is most elliptical.

The Sun is always at one focus of the planetary orbital ellipse, which means it is nearer to one side of the orbit than the other. The more eccentric the ellipse and the less nearly a circle, the closer the focus (and therefore the Sun) to one side than the other.

In the case of Venus, the eccentricity is 0.0068. (See Chapter 8 for an explanation of "eccentricity.") This means that the Sun is 66,750,000 miles from one side of the orbit and 67,650,000 miles from the other side. These distances are, respectively, the closest Venus gets to the sun ("perihelion") and the farthest it recedes from it ("aphelion"). The difference is only 900,000 miles.

When Venus is at its maximum elongation, it may happen to be at perihelion or at aphelion or anywhere in between. Naturally, if it is at perihelion it is closer to the Sun and has a slightly smaller maximum elongation than if it were at aphelion. The difference is so small, however, that it can be neglected for ordinary purposes.

Not so in Mercury's case. Its orbital eccentricity is 0.21. When Mercury is closest to the Sun it is at a disance of 28,500,000 miles. When it is farthest, it is 43,-500,000. Aphelion distance is half again as far as perihelion.

That means there is a tremendous difference between a maximum elongation when Mercury happens to be at aphelion and one when it happens to be at perihelion. The figure of 27°, given above, is an elongation at aphelion and represents the greatest of all possible maximum elongations.

At perihelion, the maximum elongation is just under 18°. At these times, Mercury hugs the horizon and sets just an hour and ten minutes after sunset or rises an hour and ten minutes before sunrise.

And mind you, these viewing times of 1 hour at some times and 2 hours at other times are only at the moment of greatest elongation, say two or three nights in every 3-month period. At all other times, Mercury is closer to the Sun and is viewable for smaller periods.

We can summarize by saying that Mercury is hardly ever visible when it is truly dark. (Astronomers use special instruments to observe it in the daylight but that is of no help to a naked-eye observer like an ancient Greek or a modern me.)

What's more, Mercury has other disadvantages as a viewable object. To be sure, it is closer to the Sun and is more brightly lit, square mile for square mile, than are other planets; *but* it is smaller than Venus and gets less total light. Where Venus has a diameter of 7,550 miles (almost that of Earth), Mercury's diameter is only 3,030 miles. Mercury is, in fact, the smallest of the planets.

Then, too, Venus has a cloud layer that reflects some ¾ of the light that falls upon it. Mercury, on the other hand, has no atmosphere and must reflect light from its bare rock surface as the Moon does. There is every reason to suppose that Mercury, like the Moon, reflects only about ¹⁄₁₄ of the light it receives.

Finally, in the crescent stage, at which Mercury and Venus are both at their brightest, Mercury is considerably farther away from us than Venus is.

Add all these things together—that Mercury is smaller than Venus, is farther away, and reflects less light—and it is not surprising that Mercury is much the dimmer of the two. Mercury, at its very brightest, has a magnitude of —1.2 and is only ¹⁄₁₇ as bright as Venus.

It is not the least bright of the planets. Saturn has a maximum brightness of —0.4, so that when Mercury is at its best it is twice as bright as Saturn. However, Saturn can be viewed in the calm darkness of the middle of the night when it may well be high in the heavens and where it will stand out as among the twenty brightest stars or starlike objects. Mercury, on the other hand, will be seen only near the horizon in dawn or twilight, amid haze and Sun glare.

There is no question in my mind, therefore, that Mercury was the seventh planet to be discovered. I suspect, in fact, that many people today (when the horizon is generally much dirtier and the sky much hazier with the glare of artificial light than it was in centuries past) have never seen Mercury.

There is even a story that Copernicus himself never saw Mercury, though I can hardly bring myself to believe that.

There is a curiosity here that I would like to point out before some Gentle Reader points it out to me. In my discussion of the seven metals of the ancients later in this book (see Chapter 12) I point out that the metal mercury was probably the seventh and last metal to be discovered by the ancients. And here I say that the planet Mercury was the seventh and last planet to be discovered, so that I entitle this chapter "The Seventh Planet."

Furthermore, it is known that the metal mercury was named for the planet Mercury. Is it entirely a coincidence that the seventh metal was named for the seventh planet? Or did the alchemists know more than is generally thought, and did they know the order of discovery of metals and planets and arrange the names to correspond?

I know there is a strong tendency on the part of mystics to believe in the "wisdom of the ancients" but I don't think they were all that wise. The correspondence between seventh metal and seventh planet is a coincidence, in my opinion. After all, there *are* such things as pure coincidence, and this is an example.

How long does it take for Mercury and Venus to go around the Sun? At first blush, it might seem that one need only count the days between one inferior conjunction and the next. The planet will then have started at the point between ourselves and the Sun, gone all around, and come back to the point between ourselves and the Sun.

It takes Mercury 116 days to do this and Venus 584 days. This is the "synodic period of revolution."

It is, however, a strictly Earth-related period. After all, the Earth is moving around the Sun, too. If Mercury starts at a certain point directly between ourselves and the Sun, then moves around the Sun back to the same point (recognized by its position relative to the stars), it will turn out that the Earth is no longer where it was. It has gone its own way and, since Mercury left, has completed about one-quarter of its own revolution.

Mercury must continue turning until it catches up with Earth and gets in between it and the Sun again. Mercury, which is so close to the Sun, moves faster than any other planet relative to it (30 miles per second, as compared with our own 18.5 miles per second). It does not take very long to catch up with us, therefore, and does so in about a month.

If we calculate how long it takes Mercury to return to the original spot relative to the stars, regardless of our own motion in the interim, then it turns out that Mercury revolves about the Sun in 88 days. This is the "sidereal period of revolution" and is the usual period we speak of as the "length of Mercury's year."

The situation is more extreme in the case of Venus. In the first place, Venus is farther from the Sun than Mercury is and must make a longer sweep to go about it. It moves more slowly than Mercury does (only 22 miles per second) and takes a much longer time to make the greater sweep.

By the time it returns to its original point relative to the stars, Earth has had a long time to move on its orbit and has gone three-fifths of a revolution.

Venus does not move much more quickly than the Earth does and it can gain on us only slowly. It must make a second full turn about the Sun and then a half-turn before it catches up.

If we count only the time it takes Venus to go around the Sun with reference to the stars and disregard the moving Earth, then the sidereal period of revolution of Venus turns out to be 225 days.

Which brings me finally to the matter of the rotation of these two planets about their axis, old view and new view, and we will continue with this subject in the next chapter.

2 The Dance of the Sun

Occasionally I receive rather depressing letters. Once, I received one that was several pages long and largely incomprehensible, but the beginning was clear enough. It objected strongly to a book of mine on astronomy.

The writer claimed that I had neglected one supremely important matter, and the implication was strong that this neglect showed my incompetence. He complained that I had spent much time describing the universe and trying to give a picture of its totality, then left out the key point. "What holds the universe up?" he demanded. "Why didn't you try to figure out what keeps it from falling?"

The proper answer would have been: "Fall where?" but that would have just gotten him angry without enlightening him. After I tried to read his explanation of what kept the universe from falling, an explanation I couldn't understand, I decided the wisest thing was not to answer at all.

But this is just one example of the type of question that puzzles mind after mind and which no explanation seems to answer properly.

There are always people who aren't satisfied that the

law of action and reaction means that a rocket can move in a vacuum. "But what are the exhaust gases pushing *against*?" they demand.

And there are some people who will never be satisfied that the Moon can be rotating if it always faces the same side to us. "If we always see the same side," they say, "then it *can't* be turning."

Naturally, there is an awful temptation (if one has an advanced case of compulsive explainitis, as I have) to try to find some analysis so clear and so irrefutable as to explain the whole thing once and for all. I have, for instance, tackled the problem of the Moon's hidden side on a number of previous occasions and now I'm going to do it yet again in a new way.

This time, though, I have an ulterior motive. I want to do it so I can continue talking about Venus and Mercury, a subject I brought up in the preceding chapter.

To begin with, each planet has two chief movements: (1) it turns, or rotates, about its axis, and (2) it turns, or revolves, about the Sun. First, let's consider the matter of rotation about the axis.

By convention, the axis (an imaginary line) defines north and south. The axis intersects the surface of the Earth at the North Pole at one end and at the South Pole at the other.

Suppose, now, we imagine ourselves high in space, exactly over the Earth's North Pole. Looking down upon the Earth, the North Pole would appear exactly at the center of the planetary circle. (Of course, only half the circle we see would be lit up by the Sun—a little more than half in the summer, a little less than half in the winter—and the rest would be in darkness, but we'll ignore that as an unimportant detail.)

From our vantage point directly above the North Pole, we would see the whole planet turning about it exactly as a wheel turns about its hub.

But which way does the planet turn? There are two possible ways, and these can be most easily described by reference to the face of a clock. We all know the way in which the hands of a clock turn. Well, the normal progression of the hands from 1 to 2 to 3 on the face of the dial is "clockwise." The opposite direction, from 3 to 2 to 1, is "counterclockwise."

As it happens, the Earth, as viewed from above the North Pole, turns counterclockwise. This counterclockwise rotation, from a view on the Earth's surface, means that our planet turns from west to east.

You can see this for yourself, if you have a mounted globe in your house (or, if you are very enthusiastic, you can find one in the local school or library). Turn the globe from west to east, and look down upon it, as it turns, from above its North Pole. You will see that it is turning counterclockwise.

However, if you continue to rotate the globe from west to east and bend down so that you can see what is happening from a view over the *South* Pole, then you will find the Earth, from that vantage point, is turning clockwise. To give meaning to the direction of turning, then, you must specify the point of view, and it is conventional always to suppose that the view is from above the North Pole.

This orientation can be applied to the solar system generally, because it happens that all the major planets are located close to the plane of the Sun's equator, and all the axes of all the planets, with the exception of Uranus, happen to be within 25° of the perpendicular to that plane.

This means that if we imagine ourselves reasonably high above the North Pole of the Earth, we are also high above the general plane of the solar system. If we then move over the Sun, we would find that one end of its axis points more or less toward us, and that the Sun moves counterclockwise about it. We would find the same thing to be true if we moved over Jupiter, Saturn, Mars, and so on.

This represents a fine display of orderliness, all this counterclockwiseness, but I warn you there are exceptions, which I will take up in the next chapter.

But what if mankind ever engages in interstellar travel

and is colonizing the planets of another Sun? Which pole should he call north and which south? To refer back to the plane of our own solar system would be inconvenient and, at times, useless. I predict that the star itself will be used as reference. That end of its axis about which it is rotating counterclockwise will be defined as its north pole. That will almost certainly define north in satisfactory fashion for its planetary system generally.

But let's get back to the Earth and the Sun. The Earth is rotating counterclockwise and we'll suppose for a moment that this is the only motion it has. We will suppose that, except for its rotation, it is motionless with respect to the Sun.

A person standing on the surface of the Earth will not feel the planet to be rotating, however. He will seem, to himself, to be standing still. Naturally, the point on which he stands is constantly changing its orientation with respect to the Sun, but that will be interpreted as caused by the fact that the Sun is moving.

As the Earth rotates from west to east, it will *appear* to an observer on its surface that the Sun moves across the sky from east to west. Because the real motion of the Earth's surface is counterclockwise, the apparent motion of the Sun is clockwise (see Figure 2).

We can measure the rate of turn as so many degrees per unit of time; say, so many degrees (°) per day where there are 360° in one complete turn.

Since clockwise and counterclockwise are rotations in opposite directions, let's arbitrarily give clockwise turns a positive sign and counterclockwise turns a negative sign.

For instance, if a planet turns counterclockwise on its axis exactly once in one day, we would say it turned —360° per day. To a person on its surface the Sun would seem to make a complete clockwise turn about the sky in one day. We could then say that the apparent motion of the Sun, resulting from the planet's rotation, is +360° per day.

We don't, however, have to pin ourselves down to any one planet moving at any one rate. We can talk about an

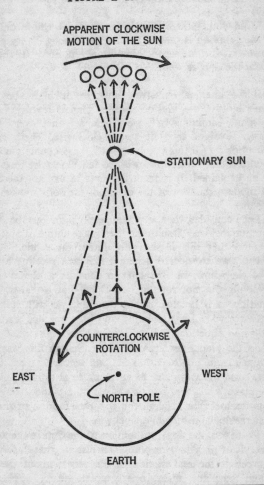

FIGURE 2 ROTATION

APPARENT CLOCKWISE
MOTION OF THE SUN

STATIONARY SUN

COUNTERCLOCKWISE
ROTATION

EAST

WEST

NORTH POLE

EARTH

arbitrary planet which happens to be turning on its axis at
a rate of $-x°$ per day. From its surface, the Sun would
seem to be moving through its sky at the rate of $+x°$ per
day.

Mind you, though, this is for a planet whose only important motion is that of turning on its axis. Every planet,
however, has a second important motion. It revolves
about the Sun as well.

If we imagine ourselves, now, to be high above the
Sun's North Pole, and observe the planets revolving about
its giant, luminous self, we would see that every major
planet revolves about it counterclockwise. There are no
exceptions. (All this regularity—all these counterclockwise
rotations and revolutions, with so few clockwise ones—has
to be explained. It is from attempts to explain them that
all modern theories of the origin of the solar system have
arisen.)

Let's consider, then, a planet revolving about the Sun in
counterclockwise fashion and, to make things simple, let's
pretend that this is the only important motion it has.
Earlier, we considered a planet that was rotating but not
revolving; now we are going to consider a planet that is
revolving but not rotating. To show that a planet is not
rotating as it revolves about the Sun, we will make the
little arrowhead, representing an observer, face always in
the same direction.

As the planet revolves about the Sun in this way, you
can see from Figure 3 that the Sun seems to move in the
sky, from the viewpoint of an observer at a fixed point on
the planet's surface. What's more, as the planet revolves in
counterclockwise fashion, the apparent motion of the Sun
that results is *also* counterclockwise.

Then, too, the apparent motion of the Sun in the sky, as
a result of planetary revolution, is just as great in terms of
degrees as the real motion of the planet about the Sun.
When the planet makes one complete turn (360°) about
the Sun, the Sun appears to make one complete turn
about the sky.

FIGURE 3 REVOLUTION

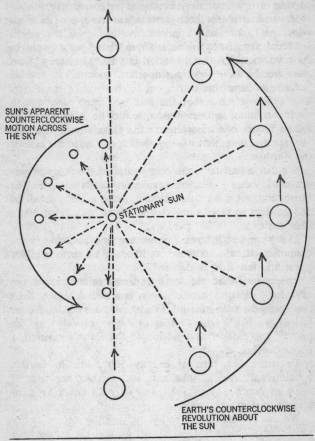

SUN'S APPARENT
COUNTERCLOCKWISE
MOTION ACROSS
THE SKY

STATIONARY SUN

EARTH'S COUNTERCLOCKWISE
REVOLUTION ABOUT
THE SUN

Suppose a planet revolves about the Sun, counterclockwise, in exactly 36 days. In 1 day, it travels $\frac{1}{36}$ of $360°$, so it is moving at a rate of $-10°$ per day. The apparent motion of the Sun is therefor carrying it, *also* counterclockwise, across the sky at $-10°$ per day.

Again, let us be general. If the planet moves about the

Sun at a rate of $-y°$ per day, then the Sun, in response, appears to move across the sky at $-y°$ per day.

If, then, a planet both rotates *and* revolves (as is always true), the Sun's apparent motion across the sky is $+x°$ per day due to the rotation and $-y°$ per day due to the revolution. The total motion is $(+x°) + (-y°)$ or, more briefly, $(x - y)°$ per day.

Let's see how this works out in the case of the Earth. The apparent motion of the Sun in Earth's sky, due to Earth's rotation *and* revolution, is such that, on the average, it makes one complete circle of the sky in one day. We can say, then, that $x - y = 360°$ per day, in the case of the Earth.

We can calculate y, the component of the Sun's apparent motion due to Earth's revolution, easily enough. The Earth completes its turn around the Sun in just about 365.26 days so that in one day it moves $\frac{1}{365.26}$ of 360°. This comes to 0.9856° per day.

If $y = -0.9856°$ per day, we can say then, that in the case of Earth, $x - 0.9856° = 360°$. If we then solve for x, we find that $x = 360.9856°$.

This means that the Sun's apparent motion about the sky, due to Earth's rotation *only,* is a little bit *more* than one complete turn of 360° in one day. Instead, it moves just about 361°, with that extra degree canceled by the part of the motion that is due to the Earth's revolution about the Sun.

To make one turn of exactly 360°, due to Earth's rotation only, would take a trifle less than one day. It takes 24 hours to turn 361°, but only 23 hours 56 minutes to turn exactly 360°.

We can see that this is so by studying the motion of the stars. Every object in the heavens beyond the atmosphere has an apparent clockwise motion across the sky in response to the Earth's counterclockwise rotation. The stars do, as well as the Sun.

The stars, however, have no progressive apparent motion across the sky in response to Earth's revolution about the Sun. In the case of the Sun, the Earth moves about it bodily so that the Sun is constantly being viewed from a changing vantage point. In the process of this bodily motion about the Sun, though, the Earth's position with respect to the very distant stars can scarcely be said to change. The stars, therefore, do not alter their apparent positions at all (except for tiny elliptical motions it takes a first-grade telescope to detect).

For stars then, the apparent motion due to rotation is $+x°$ per day, which in the case of the Earth is 361° per day, while the apparent motion due to revolution is zero. The total motion of the stars is 361° per day, and they therefore make a complete turn in the heavens (360°) in $^{360}/_{361}$ of a day, or 23 hours 56 minutes.

And if the stars are observed, the time lapse between successive crossings of the zenith meridian does indeed turn out to be 23 hours 56 minutes. This period of time is therefore the "sidereal day" (from a Latin word meaning "stars") while 24 hours is the "Solar day." The four-minute-per-day discrepancy between the sidereal day and the solar day is what makes the Sun appear to move against the background of the stars.

And what happens if the period of a planet's rotation and the period of its revolution (both counterclockwise) are equal? In that case $x = y$ and $x - y = 0°$ per day.

Thus, when a planet rotates and revolves in the same period of time, the Sun does not appear to move in the sky. It remains in the same spot constantly.

As seen from the Sun, the planet, as it revolves, would always present the same face to the luminary. In this way, the Sun would always shine on the same face from the same angle, which is equivalent to saying that the Sun does not appear to move in the sky.

The Moon presents the same face to us at all times. (The Moon revolves about the Earth so that we play the same role with respect to it that a Sun does to a planet.) This does *not* mean that the Moon is not rotating. It means that the periods of rotation and revolution are identical for the Moon.

It might seem that it is a tremendous coincidence that the periods of rotation and revolution should be equal, but, as it happens, this is not so. The gravitational influence of a large body upon a small body revolving about itself is such as to force the period of rotation and revolution into equality.

The greater the disparity in the size of the two bodies and the closer together they are, the more rapidly are the periods of rotation and revolution of the smaller body brought into equality. The Earth has succeeded in doing this to the Moon, and until very recently, it was taken for granted that the Sun had succeeded in doing so to Mercury.

Not only did gravitational theory make it seem reasonable that Mercury presented only one side to the Sun, but also observational evidence seemed to back that view.

If one side of Mercury always faces the Sun as it revolves, then only that one side is ever lit up and only that one side can ever be seen by an astronomer peering through a telescope. If Mercury is viewed over and over again at some particular point in its orbit relative to the Earth then that lit-up side ought to be seen from the same angle. Any markings on that side would be the same at each viewing.

Conversely, if the visible markings are the same every time Mercury is viewed at some particular point in its orbit, then the planet would be proved to face one side always to the Sun. It would then follow that the period of Mercury's rotation about its axis would be the same as the period of its revolution about the Sun.

The trouble is that Mercury's distance, smallness, and closeness to the Sun make its markings very difficult to observe.

Nevertheless, in the 1880's, the Italian astronomer Giovanni Virginio Schiaparelli tackled the problem. By 1890, he had seen (or thought he had seen) the same markings in the same position so often that he felt it safe to say that Mercury rotated on its axis in the time it took it to revolve about the Sun. Mercury's period of revolution is 88 days (well, 87.97 days) and its period of rotation had to be 88 days as well.

This was accepted. It fit theory, and Schiaparelli was known to be an excellent observer. For three-quarters of a century, the statement about Mercury's facing one side only to the Sun was repeated in every general astronomy book written.

But then, in 1965, radar waves were bounced off the surface of Mercury and the echoes were caught in receivers on Earth. From the nature of the echoes it is possible to tell whether the body reflecting the radar waves is rotating or not and, if rotating, how fast.

It turned out that Mercury is *not* rotating about its axis in 88 days but in 58.6 days, and this meant that it is *not* facing one side always to the Sun. Mercury's Sun side and night side, so dearly beloved by science-fiction writers, vanished into thin air.

If this is so, then how came Schiaparelli to make his mistake? Well, let's see.

If Mercury rotates on its axis in 58.6 days and does so in counterclockwise fashion, then x (the apparent motion of the Sun, due to planetary rotation) is equal to $+6.14°$ per day. The period of revolution remains 88 days, however, also counterclockwise, so that y (the apparent motion of the Sun, due to planetary revolution) is $-4.09°$ per day.

The total apparent motion of the Sun as seen from Mercury's surface, then, is $6.14 - 4.09$ or $2.05°$ per day. For the Sun to make one complete turn ($360°$) at this rate would take 176 days. To put it another way, from Mercurian noon to Mercurian noon is 176 Earth days.

Do you notice a coincidence? It turns out that 176 Earth days is exactly twice the period of Mercury's 88-day revolution. Suppose, then, that at some particular point in Mercury's orbit one particular place on Mercury's surface is directly under the Sun. Exactly two revolutions later, that same place is directly under the Sun again. And exactly two revolutions later, still again. In the in-between revolutions, that place is pointed directly away from the Sun and it is the place on the exact opposite side of the planet that gets the full blast of the Sun.

In this case, particular markings would show up on

Mercury's lighted side at particular points in its orbit in every *second* revolution.

Schiaparelli, observing Mercury, had a devil of a time making out markings, and observed most earnestly when Mercury was at aphelion and furthest from the Sun. Clouds, haze, heat quiverings must have ruined innumerable potential sightings. On a number of occasions, though, when he did make out certain markings, he imagined them to be the same, since for one thing, he expected them to be. If, on occasion, the familiar markings were missing, he felt justified in attributing it to poor visibility.

He assumed, very naturally, that if he saw particular markings so often, they were there to be seen every time, and that that meant Mercury faced one side to the Sun all the time. (I'll bet astronomers never make that mistake again.)

This peculiar form of rotation in which the planet presents first its front to the Sun, then its back, is not something that astronomers had ever predicted in advance. Now they are busily engaged in trying to find out what conditions are required to have such a situation result.

All I have said, so far, assumes that a planet's rate of rotation and revolution are constant. This is invariably true in the case of planetary rotation about its axis. It is not necessarily true for planetary revolution about the Sun.

For a planet's rate of revolution about the Sun to be constant, its orbit must be a perfect circle. If it is merely almost a perfect circle, its revolution is merely almost constant.

Mercury's orbit isn't even almost a perfect circle. At one end of its orbit, Mercury is only 28,500,000 miles from the Sun, while at the opposite end it is 43,500,000 miles from it. When nearest the Sun, Mercury moves at a speed of 35 miles per second relative to the Sun. When farthest away, it moves only at a speed of 23 miles per second.

If Mercury's period of rotation and revolution were

exactly equal, then the inconstancy of its orbital speed would prevent the Sun from remaining in one spot in the sky as seen from Mercury.

While at its closest to the Sun, Mercury would be moving so fast that y (the Sun's apparent motion due to the planet's revolution) would be considerably greater than x (the Sun's apparent motion due to the planet's rotation). This means that $x - y$ would be a negative quantity, and the Sun would be drifting west to east (clockwise) across Mercury's sky.

By the time Mercury was moving toward that part of its orbit that was farthest from the Sun, it would be moving so slowly as to fall behind the rotational period. Then y would be smaller than x, and $x - y$ would be a positive quantity. The Sun would drift west to east (clockwise).

On the whole, then, the Sun would slide first eastward, then westward, then eastward, then westward, changing direction every 44 days, but maintaining its average position unchanged. If Mercury's axis were somewhat tipped to the Sun (and the quantity of such tip, if any, is not yet certain), the Sun would seem to mark out a narrow ellipse in Mercury's sky.

But all this is only if Mercury's period of rotation were equal to its period of revolution, which it isn't. The Sun makes a complete circuit of Mercury's sky thanks to the planet's nonequal period of revolution. Imposed upon this is the back-and-forth motion produced by the planet's orbital eccentricity.

The result is a remarkable dance of the Sun of a kind no science-fiction writer has ever envisaged as far as I know.

Suppose, for instance, you were on a spot on Mercury's surface which happens to have the Sun directly overhead when it is at its closest.

You will see the Sun rise in the east, while it is actually at its farthest from Mercury. It is then a little more than twice the width of the Sun as seen from the Earth and four and a half times as hot. As it rises (which it does slowly, for there are 88 days between sunrise and sunset) it grows larger. By the time it is at zenith, it is at its closest and largest. It is then more than three times the width of the Sun as we see it and ten times as hot.

In fact, the point on Mercury's surface directly under the Sun at its largest gets extra punishment, for the Sun does not pass quickly. It is in the neighborhood of the zenith that the orbital velocity overtakes the rotational velocity and sends the Sun moving eastward again. After a while it turns and goes westward again, slipping past the zenith and continuing on to the western horizon. As it slides down through a long, long afternoon, it shrinks until it reaches its minimum size again as it sets.

On the directly opposite side of Mercury, the same thing happens. During one of Mercury's revolutions, one side gets it, during the next, the other side does—in alternation.

Even more dramatic are those places on Mercury that are 90° removed from the places that get the Sun at zenith at its largest. There, the looping motion of the Sun that arises from the nonsteady orbital velocity of Mercury takes place on the horizon, and it is there that the Sun is at maximum size.

The Sun will rise in bloated fashion in the east, rise more and more slowly, then begin to slip back and set. After a while, it rises a second time and this time it doesn't change its mind. It moves toward the zenith more and more quickly, shrinking as it goes. At zenith, it is down to minimum size.

It sweeps past zenith without turning and begins to grow again as it sinks slowly in the west. By the time it approaches the horizon, it is bloated to full size. It sets, and after a pause, it rises again as though to take a last look around and make sure all is well. Then it sets again for good.

Mercury can thus be divided into four segments, which differ as to whether the Sun is high in the sky when it is relatively large (Hot Zone) or relatively small (Cool Zone). They are arranged alternately: Hot Zone, Cool Zone, Hot Zone, Cool Zone. It should be remembered, though, that even the so-called Cool Zone is super-torrid by Earth standards.

What I have described is the dance of the Sun as seen first from the middle of the Hot Zone, then from the middle of the Cool Zone. Other places differ in the exact place in the sky (how high above the horizon) the large Sun goes through its loop-the-loop.

Nor have we even yet exhausted the peculiarities of the rotation/revolution combination. There is still Venus, to which I will turn in the next chapter.

3 Backward, Turn Backward

I rarely take vacations because I hate leaving my typewriter, but occasionally I do. Some years back, the whole family (four of us) spent a most successful three-day period in the White Mountains at the very peak of the foliage season (and if you have never seen the foliage turn in New England, you have never seen the world in living color, that's all).

We came down from a trip up Mount Washington on the cog railway (against my will, for I am no dare-devil, but I lost the vote by the narrow margin of three to one) and found that it was rapidly getting dark. That meant we had to find a motel.

My wife said, "If we turn back the way we came, we will find a whole cluster of them just a couple of miles from here."

I replied, with the kind of man-of-the-family decisiveness that made it clear there was to be no appeal, "I never turn back! We go *forward!*"

And we did go forward. We went forward through a state forest, without people or other automobiles. It got pitch dark, for there were no street lights, and for twenty-five miles we went forward through a universe of blackness, while the children got progressively more frightened and I got progressively more uneasy (what if I get a flat tire here and now!). Worst of all, we drove through the most glorious stretch of foliage color in all the world, and saw none of it.

On the whole, it was not one of my more intelligent decisions, something the family nobly forbore saying more than a mere fifty-eight times during the course of the epic drive. My only excuse was a plaintively reiterated, "But how could we go *backward?*"

There's something about going the "wrong way" or "backward" that upsets methodical, self-disciplined people like myself, and it upsets astronomers, too. They have distinct notions as to what constitutes frontward and backward in motion.

For instance, if you view the solar system from high above the Sun's north pole, then it turns out that of the nine major planets from Mercury to Pluto, exactly nine revolve about the Sun in a counterclockwise manner (as I explained in the preceding chapter).

Therefore, counterclockwise motion is the right motion, the forward motion. It is "direct motion." Clockwise motion is the wrong motion, the backward motion. It is "retrograde motion" (from Latin, meaning "backward-stepping").

To be sure, in order to judge whether a planet is revolving about the Sun counterclockwise or clockwise, from a position above the Sun's north pole, the planet's orbit should, ideally, be in the plane of the Sun's equator. This is not exactly so in actual practice. The plane of the Earth's orbit is, for instance, at an angle of 7° to the plane of the Sun's equator. The plane of the orbits of other planets is also tipped to one slight extent or another.

Fortunately, the orbital planes of the various planets are sufficiently close to one another and to that of the Sun's equator to enable us to speak of such a thing as a "general planetary orbital plane." This is very close to Earth's orbital plane, which is usually referred to as the "ecliptic." I will speak of the "ecliptic" hence-forward as the plane along which the flat structure of the solar system (the Sun and its nine major planets) lies.

The ecliptic and the plane of the Sun's equator are close enough to raise no difficulties in the matter of counterclockwise and clockwise motion. Difficulties would indeed arise if the orbital plane were tipped considerably, however.

As you can see in Figure 4, if the plane of a planetary orbit is tipped through 180°, it shifts from an equatorial orbit to what is still an equatorial orbit, but after the shift, the revolution is clockwise, not counterclockwise. Motion has become retrograde.

If the plane of the orbit is tipped through an angle of 90°, it becomes a polar orbit. Viewed from high above the Sun's north pole, the planet would be seen passing back and forth across the face of the Sun. Its orbit would be seen on edge and its revolution would be neither counterclockwise nor clockwise.

If the plane of the orbit, after being flipped 180° into clockwise motion, were flipped another 180° (360° altogether), it would be equatorial again and back into counterclockwise once more. In the process, it will have gone through another ambiguous stage at 270°.

FIGURE 4 ORBITAL TIPPING

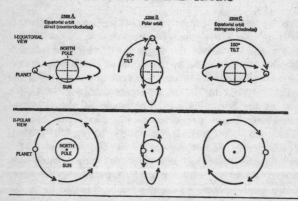

There are a couple of alternatives open to us. We can speak of "direct" movement for everything on one side of the right-angle tip and "retrograde" for everything on the other side. This is what is usually done in popular writing on astronomy, but it has its bad points. An orbit that is tipped 87° to the ecliptic would be direct and one that was tipped 93° would be retrograde, and if this were referred to *only* as direct or retrograde, respectively, there would be no indication of how close the two cases actually were or how relatively easy it might be to switch an orbit from that kind of direct to that kind of retrograde in the course of astronomic evolution.

My own recommendation would be to drop direct and retrograde as far as possible and deal only with orbital tipping. Any tip from 0° to 90° and from 270° to 360° would be direct, and anything from 90° to 270° would be retrograde.

To be sure, as far as planetary revolution is concerned, there is no fear of confusion. The greatest orbital tipping is in the case of Pluto, the orbit of which is in a plane that is tipped 17° to the ecliptic, which certainly isn't enough to confuse direct and retrograde.

Nevertheless, there are more than planets and orbital revolutions to the solar system, and there will appear some point to my argument yet.

The unanimity with which the planets revolve in direct fashion seems significant to astronomers, and it must be explained in any theory describing the origin of the solar system. The current theory has the solar system beginning with a vast cloud of dust and gas, rotating about its axis in direct fashion, and imparting that direct motion to all the parts that developed out of it. Not only do the planets all revolve directly, but the Sun rotates directly about its own axis, too.

Ideally, by this theory, there should be no retrograde motion anywhere in the solar system; yet there is. The cases of retrograde motion must be explained, if possible, without seriously affecting the general description of the origin of the solar system.

For instance, comets travel about the Sun in orbits tipped by all amounts and a number of them therefore move about the Sun in retrograde fashion.

Explanation: The comets may very well exist, to begin with, in a vast cloud spherically arranged about the Sun at a distance of a light-year or two (see Chapter 8 of *Fact and Fancy*, Doubleday, 1962). They represent the fringe of the original cloud of dust and gas, a fringe too rarefied to participate rapidly enough in the condensation of the inner portions and in the centrifugal force that spread the originally spherical cloud into a flat sheet. Consequently when some comets (possibly by the gravitational influences of other stars) are sent in toward the solar system proper from out the vast cometary sphere, they can come in from any direction and may be retrograde as easily as direct.

Next come the asteroids. A number of them have highly tipped orbital planes though actual retrograde motion is not to be found among them.

Explanation: The asteroids may have originated through an explosion of a small planet once circling in the region between Mars and Jupiter. The planet was undoubtedly revolving directly, but the explosion superimposed a second effect upon that of the original movement. Portions were sent hurling in all directions, so that some of the resulting asteroids move in fairly tilted orbits.

And what about the satellites?

Presumably, the material that was to form the planets condensed out of the fringes of the cloud that was forming the solar system, set up a whirling sub-cloud of its own. In some cases, still smaller condensations formed in its outskirts to become satellites of the central planet.

If this is so, it is to be expected that the satellites would circle the planet in, or quite near, the plane of the planetary equator and would do so directly.

Of the 32 satellites in the solar system (including Janus, a satellite of Saturn discovered in December of 1966 as a result of a queer combination of circumstances that is described in Chapter 5), no less than 20 do indeed circle their central planet in orbits that are tilted by a degree, or less, to the plane of the planetary equator. Every one of these 20 (Mars possesses 2; Jupiter, 5; Saturn, 8; and Uranus, 5) does indeed revolve directly about its planet.

But what about the other twelve? These do *not* revolve in the plane of the planetary equator and might be termed the "anomalous satellites."

The best example of an anomalous satellite is our own Moon, for its orbit is tilted considerably to our equatorial plane, the angle of tilt varying from 18.5° to 28.5° over an 18.6-year period.

Explanation: Some astronomers (and I, myself) feel that the Moon did not form out of the outskirts of the condensing cloud that was forming the Earth and therefore does not have to circle Earth in the latter's equatorial plane. Rather, the Moon may have been independently formed as a second planet in Earth's orbit and was, eventually, "captured" by Earth (see Chapter 7, *Of Time and Space and Other Things*, Doubleday, 1965).

As rather impressive evidence for that, the plane of the Moon's orbit is tilted to the ecliptic by only 5°. It is as though there was a greater impulse for it to match the Sun's equatorial plane than the Earth's, and this is exactly what is to be expected of a body that was a planet in its own right and a satellite only by later accident. Despite the tilt, the Moon's revolution about the Earth, or about the Sun, whichever way you want to view it, is direct.

The remaining anomalous satellites are distributed as follows: Jupiter, 7; Saturn, 2; and Neptune, 2. We can begin with Jupiter.

Jupiter's five innermost satellites all revolve in the planetary equatorial plane and in orbits that are nearly circular. Not so the seven outermost, which are officially known only by Roman numerals in the order of discovery. All are small, with diameters ranging from a possible 10 miles to a possible 70. All have orbits that are both markedly tilted to Jupiter's equator and markedly eccentric.

They fall into two groups. Three of them—Jupiter VI, Jupiter VII, and Jupiter X—all circle at an average distance of a little over 7,000,000 miles. (Compare this with Callisto, the outermost of Jupiter's five equatorial satellites, which is only a trifle over 1,000,000 miles from the planet.)

The remaining four—Jupiter VIII, Jupiter IX, Jupiter XI, and Jupiter XII—have orbits at average distances of from 13,000,000 to 15,000,000 miles. Jupiter VIII has the most eccentric orbit of any of the Jovian satellites. It approaches to within 8,000,000 miles of Jupiter and recedes to as far as 20,000,000.

The orbital tilt of the inner three of the anomalous satellites is about 30° or so, and their revolutionary motion is direct. The orbital tilt of the outer four, however, is something like 160° (rather close to case C in Figure 4) and their orbital motion is retrograde.

There are two questions. Why should any of the satellites revolve in retrograde fashion? And if any do, why this odd division into an inner group that is all direct and an outer that is all retrograde?

Explanation: Astronomers are generally agreed that the seven anomalous satellites of Jupiter are captured asteroids. After all, Jupiter is at the outer fringes of the asteroid belt and an occasional asteroid, passing closely by it, could be trapped in its giant gravitational field. The asteroidal nature of the satellites would account for their small size and the random nature of their capture would account for the marked tilt and eccentricity of their orbits.

It can be shown that it is easier for a planet to capture an asteroid moving in retrograde fashion than in direct. Ordinarily, there are more asteroids moving direct than retrograde and if the approach is close enough to allow the planet to pick up any asteroid, it will be direct ones that will most probably be captured (like the inner three).

Borderline approaches, however, where Jupiter's gravity just barely counterbalances the Sun's, will not yield a capture unless the asteroid can slip into a retrograde orbit, and that accounts for the outer four.

If we pass on to Saturn, we find the same situation. Of its ten known satellites, the inner eight are equatorial, and the outer two are anomalous. Whereas the outermost of the eight equatorial satellites (Hyperion) is 900,000 miles from Saturn, the ninth and tenth (Iapetus and Phoebe) are, respectively, 2,200,000 and 8,000,000 miles from Saturn. The anomalous satellites are of middle size, however, with Iapetus perhaps 800 miles across and Phoebe perhaps 200.

Again, the inner anomalous satellite has a tilt of about 10° to Saturn's equator and revolves directly. The outer one has a tilt of about 150° and is clearly retrograde.

Explanation: This would be identical to the one in the case of Jupiter.

That brings us to Neptune's two satellites, both of them anomalous. The outer one, Nereid, is a small body, perhaps 200 miles in diameter. It has an exceedingly eccentric orbit, the most eccentric orbit of any object in the solar system except for various comets. The eccentricity, 0.76, allows it to approach Neptune to within 800,000 miles and to recede to a distance of over 6,000,000. Its orbital tilt to the plane of Neptune's equator is marked but is small enough to keep its motion indisputably direct.

Explanation: Nereid, too, must be a captured asteroid, though we might wonder how it managed to get out that far. That Neptune could capture it in direct motion is probably thanks to the great distance of the Sun, which reduced the competing solar gravitational pull considerably. Even so, the extremely elliptical orbit probably indicates that Neptune just barely made the capture.

And Neptune's inner satellite, Triton? It is a large satellite, somewhat larger than our own Moon, and it circles Neptune in a nearly circular orbit at the small distance of 220,000 miles (about the distance of our own Moon from us). And, like our own Moon, it is not an equatorial satellite, but is anomalous.

Could it be that, like our Moon, Triton was an independent planet forming in Neptune's orbit? Could it be that Triton, like the Moon, was captured by its larger brother?

One fact that tends to increase the plausibility of this suggestion is that the Moon and Triton are the two largest of the satellites in terms of the mass of the planets they circle. The Moon is 1/80 the mass of the Earth, and Triton is 1/700 the mass of Neptune. All the equatorial satellites, which seem surely to have been formed as part of the planetary system to begin with and could never possibly have been independent planets, are much smaller in terms of their planets.

But if this were so, it would be nice to discover that Triton's orbital plane was at least fairly close to the ecliptic, as the Moon's is. That would give it that touch of the independent planet. Unfortunately, this is not so. The plane of Triton's orbit is tipped 220° to the ecliptic (and almost as much to the plane of Neptune's equator, which is itself tipped somewhat to the ecliptic). This means that Triton revolves in retrograde fashion about Neptune. Why should that be?

Explanation: The only one I've ever heard is that a catastrophe may have taken place. Pluto may once have circled Neptune as another satellite. Its period would have been 6.7 days (its present rotational period, see Chapter 8) and that would place it only a little farther out than Triton, which has a period of 5.9 days. Some event may have forced Pluto out of its orbit and into an independent planetary one (of unusually high eccentricity and tilt for a major planet) and the same event may have tilted Triton's orbit into its present retrograde position. What the event may have been, though, no one can suggest.

Having done with revolutions about the Sun, how about rotations about an axis? I have already said that the Sun rotates about its axis directly.

Ideally, every body in the solar system ought to do the same, with their equatorial planes all more or less coinciding with the ecliptic. To put it another way, their axes of rotation ought to be exactly perpendicular to the ecliptic. This is almost true in the case of Jupiter, which has its axis tipped only 3° from the perpendicular.

Our Moon's north pole is tipped to the ecliptic by an

even lesser extent, only 1.5° and the rotation of the Moon about its axis (and of Jupiter about its axis) is direct.

However, the Moon is the only nonplanetary body concerning the rotation of which we have clear data. The other satellites (to say nothing of asteroids) are too far away for good information on rotation. We have determined periods of rotation for Mercury and Pluto but don't have decent data on the orientation of their axis of rotation.

What about the remaining seven planets, however? I've already mentioned Jupiter; and Earth's North Pole, as we all know, is tipped through an angle of 23.5° from the perpendicular to the ecliptic. That is what gives us our seasons, makes the days and nights vary in length through the years, accounts for our climatic zones, and so on. However, the tilting is not enough to alter the fact that the Earth's rotation is direct.

Other planets have a similar polar tilt. For Mars, it is 25°; for Saturn, 27°; and for Neptune, 29°. All rotate directly, for the situation here is the same as for orbital planes. It is only when the polar tilt reaches 90° that direct motion ceases. For tilts over 90°, rotation is retrograde (see Figure 5).

But why is it that there should be an axial tilt at all?

Explanation: I don't know of any official explanation,

FIGURE 5 AXIAL TIPPING

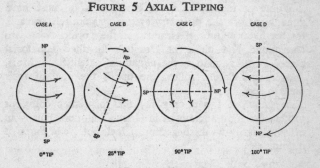

but a thought has occurred to me. As the planets form out of the original cloud of gas and dust, they would eventually form several large chunks which would, for a while, be a "multiple-planet." Circling in complex orbits they might eventually come together, striking each other off center. Might it not be that the random coming together might end by imparting a bit of "English" to the direction of rotation which would be reflected in the axial tilt? Or is this mechanically impossible? Unfortunately, I don't know.

And even if this is so, the real surprise is the close correspondence between the polar tilts of four planets: Earth, Mars, Saturn, and Neptune. Can this be pure coincidence and nothing more?

The planet Uranus has a very odd rotation; for the tilt of Uranus's north pole is 98°. This is past the 90° mark so that people (even astronomers) usually say that Uranus has a retrograde rotation. But it is not very retrograde at all, and it is misleading to call it that. If you look at Figure 5, Uranus's manner of rotation is close to that of C. (Jupiter's corresponds roughly to case A and that of Earth, Mars, Saturn, and Neptune, roughly to case B.)

Uranus actually rolls on its side, so to speak, and is not clearly direct or clearly retrograde (and it is this I had in mind when I argued against indiscriminate use of these terms early in the chapter).

Whatever managed to tilt Uranus's axis, must have done so before the planetary system was entirely formed, for the five satellites of Uranus had their orbits twisted to correspond and circle in Uranus's far-tilted equatorial plane. For that matter, the equatorial satellites of Mars, Jupiter, and Saturn follow the tip of the pole of their planets, too.

Explanation: I can only say that off-center collisions produced a right-angle twist in Uranus's motion. This is an extraordinarily weak suggestion, I feel, but I can think of nothing else.

And now we come to Venus. Until recently, nothing

was known about Venus's period of rotation or of its axial tilt—absolutely nothing. One simply couldn't see through the clouds, and the clouds themselves had no markings that could be followed around the planet as it rotated.

There were guesses, to be sure, and mistaken observations. The period of Venus's rotation was suggested as anything from 24 hours (like that of the Earth) to 225 days (its period of revolution). The consensus was that Venus's period of rotation was equal to its period of revolution and that, like Mercury, it faced one side forever to the Sun.

But then radar measurements were taken in 1964 and these showed that Venus did *not* face only one side to the Sun, any more than Mercury did (see the preceding chapter). Venus's period of rotation turned out to be 243 days, a little *longer* than its orbital period. It was the first case ever discovered of a body rotating about its axis in a period longer than that in which it revolved about a central body.

What's more, Venus's north pole was tilted at an angle of 177°, so that its case was very much like that of case D in Figure 5. This meant that Venus's rotation was retrograde—not fake retrograde, as in the case of Uranus, but real retrograde.

Obviously, we must ask why. How can Venus, in the process of formation, have been tipped over completely? Why should it be standing on its head? The notion of off-center collisions in the final stage of planet-making might account for a 25° tilt; it might just possibly account for a 90° tilt—but a complete 180° tilt?

Astronomers are bound to look for something else. Perhaps the effect of some other body in the solar system so influenced Venus as to cause it to take on this odd rotation. The natural body to suspect is the Sun, for it is fairly close to Venus and its gravitational effect would surely swamp all others.

If it is the Sun, though, then Venus ought to have a day that fits closely the period of its revolution. If it faces one side always to the Sun, then 1 Venus-rotation would equal 1 Venus-revolution. Or there might be something like the case of Mercury where 1 Mercury-revolution is equal to 2 Mercury-days (from noon to noon).

Let's check the length of the Venus-day, then, from noon to noon, according to the method described in the preceding chapter. With a 243-day rotation about its axis, the apparent daily motion of the Sun in the sky of Venus as a result of that rotation is 360°/243 or 1.48°. Since the motion is retrograde, we must have a negative sign and make that —1.48°.

Since there is a 225-day revolution about the Sun, the apparent daily motion of the Sun in the sky of Venus as a result of that revolution is 360°/225 or 1.6°. A direct revolution must also involve a negative sign so we make that —1.6°.

Adding the two figures, —1.48° and —1.6°, we have —3.08°. This means the Sun moves from west to east (if it were east to west as with us there would have been a positive sign, +3.08°) a little over 3° each Earth-day.

The Sun would make a complete circuit of Venus's sky, from noon to the next noon in 360°/—3.08° or —117 days, where the negative sign still stands for a west-to-east motion of the Sun. Consequently, we end by saying that 1 Venus-day is equal to 117 Earth-days (with the Sun moving in opposite directions in the two cases).

This means that the Venus-year is 1.835 Venus-days long, which is not a very even figure.

But consider this. The synodic period of Venus is 584 days long. This is the period (see Chapter 1) between successive moments when Venus is exactly between Earth and the Sun.

But then, the synodic period of Venus is almost exactly 5 Venus-days long. This means that every time Venus is exactly between us and the Sun, the same side of Venus, exactly the same side, faces us. It is as though Venus's period of rotation is linked to Earth somehow!

But how can that be? How can Earth's puny gravitation have an effect on Venus that would supersede the Sun's giant pull? Surely this is impossible?

Explanation: Astronomers are trying to figure this out, but I would like to advance my own theory. I wonder if

Venus's rotational period has, perhaps, not yet reached equilibrium; if it has not yet reached a final value. Perhaps, Venus's rotational period is slowly speeding up under the Sun's influence and will, some day, reach a period of 225 days *retrograde*. Its period of rotation will then be equal to its period of revolution, but in the opposite direction. The result would then be that there would be exactly 2 Venus-days in 1 Venus-year. This would be just the case that exists on Mercury, except that Venus would do it by retrograde rotation and Mercury by direct rotation.

As for the apparent connection between Venus's rotation and its synodic period—well, there *are* coincidences in the universe, and I think this is one of them. The thing to do is to continue measuring Venus's rotation as accurately as possible and see if there is a slow but steady speeding up of that rotation over the years.

And if there is—well, you heard it here first.

4 Little Lost Satellite

The older one gets, the more one tends to reminisce, I suppose. This year, as it happens, I celebrate (if that's the word I want) my thirtieth anniversary as a professional writer—something quite unbelievable to me since it seems to me I am not very much more than thirty years old.

To emphasize this fact, I have already had to renew the copyright of an even dozen of my early science-fiction stories, and when one begins renewing copyrights, it becomes necessary to face it—late youth is upon the doorstep.

So my mind keeps turning back, more than it used to, to the first science-fiction story I ever sold—back in October of 1938. The receipt of the letter of acceptance, with enclosed check, from Ray Palmer of *Amazing Stories* was one of the high points of my life. I couldn't very well frame the check for I needed the money, so I framed the letter of acceptance.

The letter was written with a dead ribbon on gray paper and only a close scrutiny made it legible at all. This was good, since it kept my modesty intact. I didn't have to say a single vainglorious word. A visitor would ask, "Why have you got an empty frame on the wall?" and I would merely say, "It isn't empty." Then my visitor would automatically approach the frame and read the letter.

Best advertisement a modest fellow ever had.

The name of that first story was "Marooned off Vesta."* I've never done an essay about Vesta, which is well worth one, and I now ought to. So I will; and we will begin with the asteroids generally, for Vesta, as you undoubtedly know, is an asteroid.

The first asteroid to be discovered was, surprisingly, the largest. No, that is not a typographical error. You might well suppose that it would be natural to see the largest first, but that isn't so in this case and I will come back to it later.

This discovery took place on January 1, 1801, under circumstances described in Chapter 9 of *Of Time and Space and Other Things* (Doubleday, 1965), so I won't go into detail here.

A second asteroid was discovered in 1802, a third in 1804, and a fourth in 1807, and there it stopped. For thirty-eight years that was how matters stood, and astronomers, who had been rattled at finding four objects in what was essentially a single orbit, settled down.

But you can always count on one troublemaker and in this case it was a German astronomer, Karl Ludwig Hencke, who, in 1830, decided he would scour the heavens and see if he could find a fifth. Year after year he searched and searched and then in 1845 he found it. And in 1847, he found a sixth, while the English astronomer, John Russell Hind, found a seventh and eighth. After that, it was just a rat race.

Some 1600 asteroids are now sufficiently well known to have had their orbits plotted, and it is estimated that there are about 44,000 asteroids with diameters of more than a mile, and heaven knows how much rubble less than a mile in diameter.

---

* Included in *Asimov's Mysteries* (Doubleday, 1968) if you are curious.

But never mind the thousands. I am going to concentrate on those first four asteroids which, between 1807 and 1845, existed in lonely splendor in the consciousness of astronomers. Their names, in the order of discovery are: Ceres, Pallas, Juno, and Vesta.

As it happens, these are large bodies as asteroids go. The best figures I can find for their diameters are listed in Table 1.

TABLE 1

Asteroids	Diameter (miles)
Ceres	470
Pallas	300
Juno	120
Vesta	240

It seems quite certain that no other known asteroid has a diameter of as much as 240 miles, so Ceres, Pallas, and Vesta are the three largest asteroids in that order, pretty much beyond question. A number of asteroids, however, are very likely to have diameters between 120 and 240 miles and some have actually been estimated (with considerable uncertainty) to be in that range. However, by virtue of time of discovery and position of orbit, Juno makes a natural fourth and we can speak conveniently, if not entirely accurately, of the "Big Four."

The Big Four have perfectly ordinary orbits, firmly between those of Mars and Jupiter. Some of the details are given in Table 2.

TABLE 2

Asteroids	Distance from the Sun (millions of miles)			Eccentricity of Orbit
	average	closest (perihelion)	farthest (aphelion)	
Ceres	257	237	278	0.079
Pallas	257	197	319	0.235
Juno	247	184	310	0.258
Vesta	219	206	239	0.088

The orbit of Ceres, as you see, is nearly circular. Its eccentricity of 0.079 is less than that of Mars. What's more, its mean distance, 257,000,000 miles, is very nearly that of the average distance of all the asteroids whose orbits are known. Vesta's orbit is only slightly less circular and it is distinctly closer to the Sun than Ceres is. Vesta's farthest point from the Sun is about as far from it as is Ceres's closest point. As for the period of revolution of the Big Four, that is in Table 3.

TABLE 3

Asteroids	Period of Revolution	
	(days)	(years)
Ceres	1680	4.60
Pallas	1686	4.61
Juno	1593	4.36
Vesta	1325	3.63

Pallas and Juno, from the data given in Table 2, seem to have nearly identical orbits. This may make it sound as if these two, particularly, are too close and are going to collide one of these days. Not at all.

The orbital diagrams you usually see in astronomy texts are two-dimensional projections of a three-dimensional reality. Orbits are tipped by different amounts to the plane of Earth's orbit (the ecliptic), as shown in Table 4.

TABLE 4

Asteroids	Tilt to Ecliptic
	(degrees)
Ceres	10.6
Pallas	34.8
Juno	13.0
Vesta	7.1

Where, in two dimensions, the orbits seem to cross, a three-dimensional model would show one crossing far above or far below another. There is no danger of collision in the foreseeable future.

It is customary to consider the asteroids as small fry and dismiss them. Indeed, the most nearly proper name given them by astronomers is "minor planets." But that is a purely anthropocentric classification. An inhabitant of Jupiter might, after all, and with considerable justification, list just four planets (Jupiter, Saturn, Uranus, and Neptune) and put everything else, from Earth on down, into the "minor planet" classification.

So let's try to consider the Big Four, at least, as objects worthy of individual consideration and see what we can find out about them.

Some have estimated that the total mass of all the asteroids is about ⅟₈₀₀ that of the Earth, or about ⅟₁₀ that of the Moon. This is only a rough guess, of course, but let's use it. This would mean that the total mass of the asteroids would be about 8,500,000,000,000,000,000 tons. The individual masses of the Big Four (also a rough guess) are as shown in table 5.

TABLE 5

Asteroids	Mass (tons)	Per Cent Total Asteroid Mass
Ceres	850,000,000,000,000,000	10.0
Pallas	220,000,000,000,000,000	2.6
Juno	14,000,000,000,000,000	0.2
Vesta	110,000,000,000,000,000	1.3

From the standpoint of mass, then, it would really seem as though there is only a "Big One"—Ceres. It is four times as massive as the second largest satellite and, within its own small sphere, it contains one-tenth of all the asteroidal matter. Together, the Big Four make up one-seventh of all the asteroidal matter.

Suppose, next, that we had reached the Big Four and decided to set up a base on one of them. What would things be like at such a station? First, how far away would the horizon seem?

As it happens, the distance of the horizon from some given low height above the surface is proportional to the square root of the diameter of the planet. Since the diameter of Ceres is just about $\frac{1}{17}$ that of the Earth, the distance of the horizon on Ceres would be about $\frac{1}{4.1}$ that of the horizon on Earth.

If we were standing on the surface of a large flat plain on Earth and were of average height so that our eyes were 5½ feet above ground, the horizon would be about 16,000 feet away (just over 3 miles). Supposing the Big Four to be smooth spheres, the horizon distance on them would be as shown in Table 6.

TABLE 6

Asteroids	Distance of Horizon (feet)
Ceres	4000
Pallas	3100
Juno	2000
Vesta	2800

It seems to me this is an important point. The dome of the sky would come down and meet the ground on Ceres, along a circle much closer to your eye than it does on Earth.

What's more, on Earth, the presence of an atmosphere dims objects on the horizon and turns them bluish. (We can see the tops of hills and mountains much farther than 16,000 feet away, of course, and they are all the mistier and bluer for that.) We use this mistiness as a way of unconsciously estimating distance. When the air is very clear and distant objects seem sharper than usual, we automatically think they are closer than they really are.

On Ceres, where there is no atmosphere, the objects on the horizon would be sharply outlined. Even the tops of crags farther than 4000 feet away would be sharp. We would therefore estimate the horizon on Ceres to be closer than it actually is. It seems to me certain, then, that the men establishing a base on Ceres (and even more so on the other asteroids) would be subject to claustrophobic uneasiness. Some, who are used to wide-open spaces, might not be able to make it; perhaps the asteroids had better be staffed by city boys.

In another way, the Big Four are not so small after all. What about their surface area?

TABLE 7

Asteroids	Surface Area (square miles)
Ceres	700,000
Pallas	280,000
Juno	45,000
Vesta	180,000

The surface area of Ceres is just about as large as that of Alaska plus California. Even the surface area of Juno, the smallest of the four, is equal to that of New York State, and the total area of all four is equal to one-third that of the fifty United States. There is plenty of room to explore on the Big Four and, for that matter, plenty of room to get hopelessly lost in.

The question of surface gravity may make the Big Four seem small again. If two spherical bodies are of equal density then the surface gravity is proportional to the diameter. If we assume that the Big Four are essentially rocky in character, then they probably have the

TABLE 8

Asteroids	Surface Gravity (per cent of Earth's)	A 180-Pound Man Will Weigh (pounds)
Ceres	3.5	6.3
Pallas	2.2	4.0
Juno	0.9	1.6
Vesta	1.8	3.2

same density as the Moon. The surface gravity of the Moon (2160 miles in diameter) is 0.16 that of the Earth. In that case, we can prepare Table 8.

This looks like a set of feeble grasps indeed. Is it possible that a person, making a careless move, might shoot upward and away from the satellite altogether? Would the feeble gravity fail to hold him and would he be lost in space? To test the danger of that, it is only necessary to calculate the escape velocity.

TABLE 9

Asteroids	Escape Velocity	
	(miles/second)	(mile/hour)
Ceres	0.33	1200
Pallas	0.21	750
Juno	0.08	300
Vesta	0.16	600

These values are not high compared to Earth's escape velocity, which is 6.98 miles per second, or 25,000 miles an hour. Still even on Juno, the baby of the four, one would have to move at a speed of 300 miles per hour to lift off the asteroid. You are certainly not going to jump upward at that speed. You are not even going to drive a ground vehicle at that speed. You will therefore be held by the asteroid, and the gravitational force, small though it may seem, will do its essential job.*

But now let's raise the question of the order of discovery. I said, at the start of the chapter, that it was surprising that the first satellite to be discovered was the largest.

* Since this article was first written, a letter from Thomas McNelly of Cornell gave me some calculations as to the influence of the centrifugal effect introduced by the possible rotational periods of the asteroids. At the equator, the gravitational pull might be reduced from 5 to 10 per cent, and the escape velocity from 30 to 40 per cent. You would still remain securely on the asteroid but by a narrower margin than I had originally assumed.

The reason that is surprising is that you would expect the first satellite to be discovered to be the *brightest,* and the largest is not necessarily the brightest.

Being large helps, of course. All other things being equal, a large body catches more sunlight than a small one and is brighter. But are all other things equal? Two factors in particular may affect matters: distance and albedo.

It is clear that the farther from the Sun an asteroid is, the less light it catches and the less bright it is. A small asteroid near the orbit of Mars ought to be brighter than a considerably larger asteroid near the orbit of Jupiter, and the smaller one ought then to be discovered sooner than the larger.

But as it happens, the Big Four are moderately close to us as asteroid distances go, and there are no reasonably large ones that are markedly closer than they are. It is not to be expected then that any asteroid will be discovered sooner than the Big Four merely because of a distance difference.

Furthermore, the Big Four themselves are not at markedly different distances; not enough different, at any rate, to overcome the size differential. Eliminate distance, then.

What about albedo?

Albedo is the fraction of the light, received by a planet, which is then reflected. Thus, if a planet reflects one-fifth of the light it receives from the Sun, it has an albedo of 0.2

As it happens, the solid rock of a planetary surface is a poor reflector of light. The Moon, which has a surface that is all rock and that is all exposed to direct sunlight (since there is no atmosphere present), reflects less than $1/16$ of the light it receives. Its albedo is 0.06, and the same is true for Mercury.

A cloudy atmosphere is much better at reflecting light. Mars, with a thin atmosphere and an occasional thin cloud, has an albedo of 0.15, two and a half times that of Mercury and the Moon.

Planets with thicker atmospheres reflect an even larger

fraction of the light they receive. The albedo of the Earth is 0.40, while that of the outer planets approaches the 0.50 mark. Venus's clouds do best of all for some reason, and its albedo is about 0.70.

But with respect to the asteroids, the albedo should raise no problem. It passes the bounds of belief that even the largest asteroid should be able to retain an atmosphere. If we consider the asteroids, generally, as composed of rock, all should have an albedo of 0.06. (Indeed, asteroids other than the Big Four have their diameters calculated from their known distance and brightness by assuming this albedo.)

Therefore, we can (it would seem) eliminate the albedo as a factor. We can expect the brightness of the Big Four to decrease with size, and therefore we can expect that chances are the Big Four would have been discovered in order of size.

The apparent brightness of an astronomical body is measured in magnitude, which is a logarithmic scale. That is, a body of magnitude 7 is 2.5 times as bright as one of magnitude 8 and is 2.5 × 2.5 or 6.25 times as bright as one of magnitude 9, and so on. (Notice that the higher the magnitude, the less the brightness.)

If we consider the first three asteroids to be discovered, all this works out neatly, as seen in Table 10.

TABLE 10

Asteroids	Diameter (miles)	Magnitude (closest to Earth)	Brightness (Juno=1)	Year Discovered
Ceres	470	7.4	3.3	1801
Pallas	300	8.0	1.9	1802
Juno	120	8.7	1.0	1804

Ceres, the brightest of the three, is too dim, at its brightest, to be seen by the unaided eye, but even a small telescope will show it, so you might suppose it only right that it was first discovered; then Pallas, which is over half as bright as Ceres; and then Juno, which is over half as bright as Pallas.

But Vesta is larger than Juno, and should be brighter. Why was it discovered only in 1807, three years after Juno? It can't be distance because it is even closer than Juno. Perhaps then, it is albedo. Perhaps, for some reason, Vesta is composed of darker rock and is dimmer than Juno despite the larger size of the former.

Yet that is not so either. In fact, Vesta's magnitude, far from being greater than that of Juno, is not only less than Juno's but less than that of Pallas and even of Ceres. Vesta has a magnitude, at closest approach, of 6.5, which means that it is no less than 7.5 times as bright as Juno. It is even 2.3 times as bright as Ceres.

To put it most sharply, Vesta is the brightest of all the asteroids and at its brightest can just barely be made out on a dark, moonless night by someone with excellent eyes. It is the only asteroid that can ever be seen with the unaided eye.

Why, then, did it take so long to discover Vesta?

Ceres was discovered by accident. Its discoverer wasn't looking for any planetary body and he happened to spot Ceres, the second brightest asteroid. That's reasonable enough.

Still, for six years after that, a group of astronomers searched intently for other asteroids. Why is it they found Pallas and Juno before they found the much brighter Vesta—*years* before?

But that's a minor mystery, after all. What is much more puzzling is why Vesta should be so bright.

The surface area of Vesta is just about ¼ that of Ceres. Allowing for the fact that Vesta is a trifle closer to the Sun and gets more light, it should be about ⅓ as bright as Ceres if the two were of equal albedo. The fact that Vesta is actually ⅞ as bright as Ceres means that it must have an albedo that is seven times as high. The albedo of Vesta, in fact, is possibly as high as 0.5, a value equal to that of planets with deep, thick atmospheres.

But Vesta *can't* have a deep, thick atmosphere. The only remaining alternative is that it has an icy surface;

that there are fields of ice (or possibly frozen carbon dioxide) on Vesta's surface, reflecting, with considerable efficiency, the feeble light of the distant Sun.

But if that is so, where did the ice come from? Why should there by only one asteroid out of all those thousands that is icy?

Or *is* it just one? Is it possible that there are a whole class of icy asteroids? Is it just that of the four, whose diameter we happen to have measured directly with fair accuracy, only one is icy, and that that gives us a false impression of uniqueness?

All the other asteroids have had their diameters determined on the assumption of a low albedo. Suppose a number of them have high albedos and are considerably smaller than we assume.

Perhaps an original asteroid-planet exploded, its interior forming stony asteroids (with a few nickel-irons from its core, if it had one) while its ice-encrusted surface gave birth to Vesta-type asteroids; with only Vesta itself, of that type, clearly visible.

But there are problems. Would the catastrophe leave the ice on a surface fragment intact? Wouldn't the ice be blown off or melted off by the energies released by the explosion, and distributed through space? And would the fragment, undoubtedly irregular to begin with, coalesce into the fairly spherical shape Vesta now has, with ice remaining on the surface rather than folding into the interior?

And if the explosion theory is eliminated, what is the alternative?

One suggestion is that Vesta *is* unique, because it is not an asteroid at all, but a displaced satellite. If it had been a satellite to begin with it would have been a sphere from its time of formation and it might have picked up an ice layer from the outer atmosphere of the young planet it was circling. Then, somehow, the satellite was pulled away from the planet and went wandering off, lost, in the asteroid belt.

It's a touching picture of a little lost satellite, but what would its original planet have been? The closest planets to the asteroid belt are Jupiter and Mars. Could Vesta have once been a satellite of Jupiter?

We have one piece of information that might help us decide. In 1967, careful measurements of the brightness of Vesta were made and a tiny variation with a period of 5⅓ hours was reported. Supposedly, this represents its period of rotation, for some parts of its surface may be less densely covered with ice than others. As the relatively rocky side shows, the albedo drops and with it the brightness; as the icy side shows, both go back up again; and this is repeated over and over.

Such a rotation ought to represent Vesta's original period of revolution about a planet; for if it had been a satellite, it would very likely have presented but one face to its planet and have had a rotation equal to its period of revolution, like our own Moon.

For a satellite to circle Jupiter in 5⅓ hours, it would have to be 64,000 miles from Jupiter's center, or only 20,000 miles above its visible cloud layer—much closer than Jupiter's closest present satellite.

A satellite in that position—even if it could withstand the tidal strain of Jupiter's vast gravity—couldn't possibly have been snatched away from that planet's enormous grip by any reasonable mechanism. Even if some unimaginable catastrophe had ripped away an inner-satellite-Vesta from Jupiter, how could it have done so without disturbing Jupiter's other inner satellites?

(How can we tell the other inner satellites weren't disturbed? Well, Jupiter's five innermost satellites move about it in almost perfect circles and almost exactly in the planetary equatorial plane, and that can only be true for satellites that have never been seriously disturbed from the moment of their formation.)

If Vesta were originally a satellite of Mars, the situation would not be much better. To revolve in 5⅓ hours it would have to circle Mars at a distance of 4500 miles from its center, or 2400 miles above its surface. It would be closer to the planet than either Phobos or Deimos and it could not have been abstracted from its position without disturbing Phobos and Deimos, and those two satellites have not been disturbed.

Besides, if Vesta had been abstracted from anywhere by some cosmic catastrophe, it would very likely have an eccentric orbit; but it doesn't, not particularly. And it revolves just about where a normal asteroid should. That's asking a lot of coincidence for a little lost satellite.

(On the other hand, Richard H. Weil wrote me after this essay first appeared suggesting that Vesta may be the satellite of the original unexploded asteroid-planet, a most attractive suggestion. What's more, the well-known science-fiction writer James Blish used this hypothesis in a story of his—which, alas, I had missed.)

So there is a fascinating mystery about Vesta, and because of that very fact, it may have more to tell us about the asteroids generally, and about the solar system as a whole, perhaps, than any other body between Mars and Jupiter.

When the time comes, then, that our manned spaceships head out beyond Mars at last, I want to put in a strong suggestion for the port of first call.

Vesta, please!

B
THE
OUTER
SYSTEM

5 Little Found Satellite

One week ago (as I write this) I visited Brandeis University with a friend to look over a tremendous exhibit of old Bibles which they had on view. We stopped at one fifteenth-century Bible, published by Spanish Jews, which was open to the seventh chapter of Isaiah. This contained the verse which, in the King James version, reads in part, "Behold a virgin shall conceive, and bear a son."

The Hebrew word *almah,* which is translated "virgin" in the King James, was not translated in the Spanish Bible we were looking at. It was merely transliterated and left "almah" in Latin letters, a fact pointed out in the information card that accompanied the exhibit.

My friend wondered why that was so and I explained that *almah* didn't really mean "virgin." (In the Revised Standard Version, the verse is translated, "Behold, a young woman shall conceive and bear a son.") Yet it would have been terribly dangerous for Spanish Jews of that period to translate correctly and seem to be denying the virgin birth, so they evaded the issue by leaving the word untranslated.

I'm afraid my enthusiasm caught up with me at this

point, as it so often does when I am trapped into explaining something. My voice reached its normal window-rattling pitch and I grew oblivious to my surroundings. What's more, I went on to explain my own theories (at considerable length) as to the significance of this chapter of Isaiah and grew eloquent indeed. Finally, after what must have seemed an eternity to my friend, I ran down and passed on to other items in the exhibit. The sequel was told me, with obvious delight, by my friend later.

It seems a Pinkerton guard at the exhibit (which was valued at $5,500,000) had been listening to me. After I passed on, the guard said to my friend, "What makes him such an expert?"* and my friend, with greater faith in the weight of my name than I myself have, said, portentously, "Do you know who he is?"

The guard thought a little and said, "God?"

Oh, well, I suppose that sort of thing is an occupational hazard for the professional explainer and I'm not going to let it rattle me. I will continue to practice my profession though I will try to lower my voice next time and, perhaps, somewhat subdue my general air of divine authority.

With that in mind let's take up the subject of the planet Saturn.

Saturn was known as one of the planets from the beginning of astronomical records. Its magnitude at its brightest is −0.4, and it is then brighter than any star except Canopus and Sirius, so it is easily seen, even though it is dimmer than any of the other planets at *their* brightest (see Chapter 1).

What made it remarkable, as far as the ancient astronomers were concerned, was that it moved more slowly against the background of the fixed stars than did any other planet. The Moon made a complete circle of the sky

* You may be wondering the same thing. I'm not, really, but I have written a rather big book about the Bible, the first volume of which was published by Doubleday in October, 1968, so I'm up on the subject. The title of the book, in case you can hardly wait to buy it, is *Asimov's Guide to the Bible*. Need I explain that the title is the publisher's and not mine?

in a month; the Sun in 1 year; Jupiter in 12 years. Saturn, however, did not complete a circle of the sky until 29.5 years had passed.

The Greeks interpreted this slow movement to signify that Saturn was farther from the Earth than was any of the other planets known to them. In other words, it seemed to move so slowly because it was so distant. That would also account for the fact that it was dimmer than the other planets.

In this, the Greeks turned out to be correct Of course, they had no way of knowing the actual distance of Saturn, but, on the average, it is about 887,000,000 miles. It also, as a matter of fact, moves more slowly about the Sun than any of the other visible planets do. Whereas Mercury's average orbital speed is 29.8 miles per second, and Earth's is 18.5 miles per second, Saturn moves along at a mere 6.0 miles per second.

The Greeks called the planet "Cronos." The myths concerning Cronos are rather unpalatable. He was the son of Ouranos (Uranus), the god of the sky, and Gaia (Gaea), the goddess of the earth. His parents had given birth to a series of monsters whom Ouranos in horror had imprisoned in hell. Gaia angrily encouraged her non-monstrous son, Cronos, to take vengeance.

Armed with a sharp sickle, Cronos crept up on his father when the latter was asleep, castrated him, and took over the rule of the universe.

But the crime had its aftermath. Cronos feared that his children would serve him as he had served his father. Therefore, each time his wife, Rhea, bore a child, Cronos would swallow it. Finally, when Rhea bore Zeus, she deceived Cronos by giving him a stone dressed in swaddling clothes. Zeus was reared in secret, and when he matured, he warred against Cronos, defeated him, made him disgorge his other children, and then replaced him as ruler of the universe.

The planet we call Jupiter was called Zeus by the Greeks. It circles the heavens in just under 12 years. Every 20 years, it catches up to the planet Cronos, and passes it. Because Cronos moves so slowly, it is fitting to name it for an old, old god. (And the slow, heavy motion gives us our word "saturnine.") Because Zeus is forever overtaking and passing it, the two seem forever to be reenacting the old myth of Zeus replacing his father on the universal throne.

The Romans identified their god, Saturn, as agricultural deity, with Cronos. (And perhaps they were right to do so, for Cronos himself may have been an agricultural deity to begin with. The sickle he used is an agricultural implement and the castration myth may refer, symbolically, to the harvesting of grain, where the fertile ears are cut off.)

There is a strong tendency to confuse Cronos with *Chronos*, which *is* a Greek word and which means "time." For that reason, the aged Cronos with his sickle is used to represent "Father Time." It is an excellent representation, as it happens, for the relentless sickle mowing down everything is a chilling picture of what time does, but we pedants consider it wrong just the same.

In July, 1610, the Italian astronomer Galileo Galilei turned his telescope on Saturn. He had already discovered the mountains on the Moon, four satellites of Jupiter, sports on the Sun, and new stars everywhere. What could he find out about Saturn?

His telescope was, of course, a primitive one, and Saturn was far away. Galileo couldn't make out clearly what it was he saw, but he saw *something* odd. Saturn wasn't just a round ball, as Jupiter was; there was a bright something or other on either side. Galileo thought they might be a pair of subsidiary bodies, large twin satellites, one on either side.

In 1612, when he had a chance to return to the study of Saturn, he found the satellites (if that was what they were) had disappeared. He saw nothing but a round ball, like Jupiter, only smaller.

Galileo was quite upset. After all, the telescope was a brand-new instrument, and there were not wanting opponents who were insisting that what it made visible were merely illusions, sent by the devil to rouse doubt concerning the divinely inspired Biblical account of the universe.

If the telescope showed Saturn to be a triple body at some times and a single body at others, that would indeed be a triumph for reaction.

Galileo avoided looking at Saturn after that, and in a letter to a friend, asked, petulantly, "Does Saturn swallow its children?"

(I was once asked quite seriously whether the Greeks might have been able, somehow, to see Saturn plainly enough to make this out, and named the planet accordingly. Of course not! The Greeks had no telescope and it is as certain as anything can be that they never saw Saturn as anything but a point of light. The matter is a coincidence, though a particularly interesting one.)

Nearly half a century later the Dutch astronomer Christian Huygens, with better telescopes at his disposal, began to study Saturn. In 1655 he discovered it had a satellite and named it Titan.

By the next year, he had puzzled out what it was that had so upset Galileo. His telescope wasn't quite good enough to make it out with indisputable clarity, but he, too, could see something extending out on either side of Saturn.

Call them a pair of satellites, as Galileo had done. If that were so, then they would have to revolve about Saturn.

That was a slightly risky deduction, for this was before Newton had advanced his law of universal gravitation which gave a sound theoretical basis for maintaining that a satellite *must* revolve about its planet. Still, the Moon revolved about the Earth, the four satellites of Jupiter each revolved about Jupiter, and Titan revolved about Saturn. Why not suppose that Saturn's twin satellites followed the general rule?

If these twin satellites revolved about Saturn, their appearance should alter from night to night. Eventually one should be behind Saturn and one in front so that both would be invisible.

Could this be what accounted for the fact that Galileo didn't see them in 1612? No! The visibility and invisibility periods should alternate at intervals of a very few days or even hours if the objects were twin satellites, and not at intervals of several years.

The only way Huygens could account for the fact that the appearance of the twin satellites remained unaltered for night after night was to suppose that there were a number of satellites so that there were always some in front and back of the planet and some to either side. Indeed, to make the appearance utterly constant, the satellites ought to be in the form of a ring about the planet.

But then how account for the fact that the appearance did *slowly* change over the years and that the satellites disappeared utterly from time to time?

Huygens reasoned that the ring might be a thin one which was tipped to the plane of the ecliptic (to that of Earth's orbit, in other words). This is so, and the angle of inclination is now known to be 26° 44'.

When Saturn is on one side of its orbit, we look down on the rings and see them clearly in front and to either side. When Saturn is on the other side of the orbit, we look up on the rings and again see them clearly.

If we could watch Saturn every night as it swings from one end of its orbit to the other, we would watch the tipping of the ring begin at maximum-down to maximum-up over a period of 14¾ years as the planet goes from one end to the other. In another 14¾ years it goes back to its initial point and the ring shifts from maximum-up to maximum-down.

(Actually, the tilt of the rings does not shift. It remains constant with reference to the stars. This constant-tilt-with-reference-to-the-stars combines with Saturn's motion around the Sun to produce an apparent oscillation-of-tilt with respect to the Sun, to Earth. This is not easy to see in the mind alone, so maybe Figure 6 will help.)

Midway between the maximum-up and maximum-down, there must come a time when the rings are seen exactly edge-on. The same thing happens midway between the maximum-down and the maximum-up. This means that every 14¾ years, the rings are seen edge-on. If they are a thin, flat structure, and our telescopes are not big enough

FIGURE 6 THE TILTING OF THE RINGS

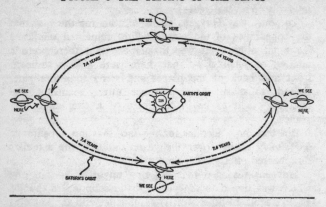

to magnify the rings to the point where the thickness is perceptible—they will seem to disappear. That is what happened to Galileo.

Huygens put his discovery into a few Latin words and then scrambled the letters by putting them in alphabetical order. He wanted to preserve priority if he turned out to be right, while reserving the privilege to withdraw without embarrassment if he were wrong, as did Galileo in an earlier case (see Chapter 1).

By 1659, he was convinced he was right and he published the Latin sentence, which read "Annulo cingitur, tenui plano, nusquam cohaerente, ad ecliptam inclinato." Translated freely, it says that Saturn is "surrounded by a thin, flat ring, not touching it anywhere, and tilted to the ecliptic."

Huygens is not to be blamed for his caution. There was simply nothing like Saturn's ring in the heavens, *and there still isn't*. It is absolutely unique among all the heavenly features we can see.

What is the ring? It appears to be a thin, flat, solid structure. Is that all?

Huygens's contemporary, the Italian-French astronomer Jean Dominique Cassini, is the next great Saturnian. In

1671 and 1672, he discovered two satellites (Iapetus and Rhea) and in 1684 two more (Dione and Tethys) so that at that time five Saturnian satellites were known altogether.

And in 1675, he published his contribution to knowledge concerning the ring. He detected a dark line curving around it about a third of the way in from the outer edge. Huygens thought it was a dark marking on a solid ring. Cassini thought it was a separation and that there were really two rings, one outside the other. The outer one we can call Ring A and the inner Ring B.

Cassini turned out to be right, and the dark marking he discovered is still called "Cassini's division." As a matter of fact there are other divisions, too, though none are so broad and easy to see as Cassini's. Anyway, since 1675, we speak of the "rings" of Saturn, not the "ring."

New satellites continued to turn up. In 1789, the German-English astronomer William Herschel discovered a sixth and seventh satellite (Mimas and Enceladus). He also studied the rotation of Saturn by following spots on its surface. He found that Saturn rotated in 10¼ hours and that its axis of rotation was perpendicular to the plane of the rings. The rings, in other words, circle Saturn along the line of its equator. The inner satellites also revolve exactly in the plane of Saturn's equator.

The rings cast a shadow on the planet they circle so it is easy to suppose they are solid and continuous. A strong point in favor of an alternate possibility, however, came with the work of the American astronomer William Cranch Bond.

In 1848, Bond detected an eighth satellite of Saturn (Hyperion). He made the discovery on September 16; the English astronomer, William Lassell, independently made it on September 18. Two days is two days—Bond gets the credit.

Next, in 1849, Bond and his son, George Philips Bond, observed that the rings of Saturn extended closer to the planet than had been reported. There was an innermost ring that was dimmer than the rest and through which stars could be seen. That part (Ring C) certainly couldn't be continuous. The Bonds made their report on November 15. Lassell made an independent report on December 3 and lost out again, poor fellow. (See Figure 7.)

So much for observation. There was theory, in addition, and that, too, pointed to rings made up of discontinuous particles.

The rings spread out widely in space and were under the influence of a gravitational field that varied in intensity with distance. The innermost boundary of the ring system is only 44,000 miles from Saturn's center (and only 7000 miles above Saturn's surface), while the outermost boundary is 86,000 miles from the center of the planet.

FIGURE 7 SATURN AS SEEN FROM ABOVE ONE POLE

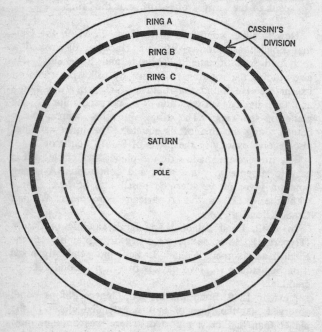

We know Saturn's mass. Newton worked it out first in 1687 from the distance of its satellites and their period of revolution, using his law of universal gravitation, and his figure has had to be corrected only slightly since. Saturn has 95 times the mass of Earth and from that we can calculate quite easily that the gravitational intensity at the innermost boundary of the rings is 0.8 and at the outermost boundary 0.2 (where the gravitational intensity at the surface of the Earth is taken as 1.0).

In other words, the outermost regions of the ring are under only a quarter the gravitational intensity of the innermost regions. This gives rise to a strong tidal effect that puts a strain on the rings inward and outward.

Then, too, under the lash of the gravitational force, the innermost sections should move at a velocity of 12½ miles per second in their revolution about Saturn, while the outermost sections should move at 10 miles per second. If the rings were solid, they would have to move in one piece with the innermost sections going more slowly than the outermost. This would put a tremendous sideways strain on the system.

In 1857, the Scottish mathematician James Clerk Maxwell made a thoroughgoing analysis of the gravitational situation and showed that the rings could not stand the strain if they were continuously solid unless they were many times as rigid as the best steel. They had to consist of separate fragments so thickly strewn as to appear continuous when viewed from the distance of Earth (not so thickly strewn in Ring C).

This has been borne out amply since. The American astronomer James Edward Keeler, in 1895, showed by spectroscopic observations that the inner portion of the rings *did* move more quickly than the outer.

Despite everything, though, we still don't know how thick the rings are. As telescopes improved and as the rings *still* remained invisible when seen edge-on, the estimates of maximum thickness decreased. At first, it was felt they had to be less than several hundred miles thick, then less than several dozen miles thick, then less than ten miles thick. Now it is generally felt their thickness must be measured in yards at the most. I have seen some estimates to the effect that they are only a foot thick.

The size of the individual particles in the ring is also not known. The thinner the rings, the smaller the particles might be expected to be. They may be no larger than so much gravel. From the efficiency with which they reflect light and from their infrared spectra, it seems the gravel may be coated with ice, or even be mainly ice.

But why should Saturn have rings at all? No other planets have them.

Well, in 1849, before Maxwell's definitive analysis, a French astronomer, E. Roche, made a more general approach and showed that when a satellite and its primary are of the same density, tidal forces will break up the satellite if it is closer to the planet's center than 2.44 times the planet's radius. This is called Roche's limit.*

There are six planets known to have satellites. If we omit Saturn, here are the figures for the distance of Roche's limit for each planet and the distance of its nearest satellite:

TABLE 11

Planet	Roche's Limit		Nearest Satellite	
	miles	radii	miles	radii
Earth	9,750	2.44	238,500	60.0
Mars	5,150	2.44	5,800	2.74
Jupiter	105,000	2.44	110,000	2.56
Uranus	37,000	2.44	77,000	5.17
Neptune	35,500	2.44	220,000	15.1

For every one of these planets, the satellite system is outside Roche's limit although, in the case of Mars and Jupiter, the innermost satellite is interestingly close to that limit.

* Actually, Roche calculated it on the assumption the satellite is fluid and has no tensile strength. Actual solid satellites can withstand tidal forces better than that and if a satellite is small enough it can remain unbroken even quite close to a planet.

For Saturn, the distance of Roche's limit is 88,500. The innermost known satellite, as of December, 1967, was Mimas, which is 120,000 miles from Saturn's center (3.44 radii). Saturn's satellite system is also safe.

The outermost edge of the rings is, on the other hand, 86,000 miles from Saturn's center, or 2.38 radii. The entire ring system, therefore, is inside Roche's limit. Either it represents a satellite which somehow managed to get too close to Saturn and which broke up, or it represents part of the original cloud of matter about Saturn which was too close to Saturn for tidal effects to allow it to coalesce into a satellite in the first place.

If the rings are a disintegrated satellite, it may have been a rather large one. I have seen one estimate which placed the total mass of the rings at one-quarter that of the Moon. If so, the original satellite would have had to have a diameter of about 1300 miles.

And what about Cassini's division? If there were particles in the division (which is about 2500 miles wide, by the way), they would revolve about Saturn in 11 hours. However, Mimas revolves in 22 hours, Enceladus in 33 hours, and Tethys in 45 hours.

A particle in Cassini's division would be pulled by Mimas from the same direction every two of the particle's revolutions, by Enceladus every three, and by Tethys every four. The same pull from the same direction would permanently slow it down and force it closer to Saturn, or speed it up and force it farther from Saturn until the new distance is such that the synchronization is destroyed. In this way, Cassini's gap is swept clear of particles.

Other gaps are likewise swept by synchronization with the nearer satellites, but nowhere is the sweeping as multiple and as efficient as in Cassini's division.

But now let's consider Saturn's satellites. In 1898, the American astronomer William Henry Pickering discovered Phoebe, Saturn's ninth satellite, and for almost seventy years after that, nothing new was added. The list of satellites, from Saturn outward, was: Mimas, Enceladus, Tethys, Dione, Rhea, Titan, Hyperion, Iapetus, and

Phoebe. It almost seemed as if that were going to be permanent.

To be sure, Pickering reported a tenth satellite in 1905 and named it Themis. He reported it at a distance of 908,000 miles from Saturn, which placed it at nearly the orbit of Hyperion. Two satellites so near the primary are not likely to be so closely spaced. Of course, Themis was reported to have a high inclination of 39° and a rather high eccentricity as well, so that it wasn't likely to collide with Hyperion. However, a high inclination and eccentricity for a satellite that close to Saturn are also unlikely. It is a suspicious satellite all around and it is not at all surprising that Themis was never seen again. Pickering must have been mistaken, that's all.

A close study of faint divisions in Saturn's rings, however, indicated that additional satellites, very close to Saturn, might exist. If so, they would be so close to the outer edge of the rings as to be lost in their glare. Then, in December, 1967, something dramatic happened. This was a time when Saturn's rings showed edge-on and disappeared, and such a once-in-fifteen-year opportunity can be important. With the rings out of sight, objects in the near vicinity of the rings might be seen.

In that December, a French astronomer, Audouin Dollfus, located a tenth satellite of Saturn, closer to the planet than Mimas, even. The little found satellite proved to be 98,000 miles from the planet's center (2.73 radii), which puts it 22,000 miles closer than Mimas, but still 10,000 safe miles outside Roche's limit and 12,000 miles beyond the outer edge of the rings. The new satellite revolves about Saturn in 18 hours.

Dollfus named it Janus after a Roman god who is pictured with a double face, one looking forward and one backward. Janus is the god of beginnings and endings, of arrivals and departures, of entrances and exits, of doors in and doors out (which is why the guardian of the doors, and, eventually, of the house generally, is called a "janitor"). In this case, the new satellite is the last to be discovered and the first on the list reading from Saturn outward, so it is "Janus."

Junus's diameter is estimated at 300 miles, which would make it the largest satellite discovered in over a hundred years. Considering how telescopes have improved in that interval, you can see that Janus would surely have been discovered long ago if it hadn't been for the rings.

Janus's invisibility is another case of Saturn (and of its rings) seeming to swallow its children. Dollfus was the Zeus that made it disgorge.

6 View From Amalthea

Several months ago I attended a preview of the motion picture *2001: A Space Odyssey* here in Boston. Against my better judgment I even got into a tuxedo for the occasion.

Perhaps the tuxedo contributed to an unwitting bit of pomposity on my part, for at one point I dissolved into a semi-irrational spasm of anger.

You see, I have included in a number of my stories something I call "The Three Laws of Robotics," of which the first is: "A robot may not harm a human being or, by inaction, allow a human being to come to harm." The laws are a purely fictional device, but they had been picked up by other writers, who take them for granted in *their* robot stories, and over the years I have come to take them very seriously indeed.

In *2001*, the most dramatic episodes involve an intelligent computer (equivalent to one of my robots) who deliberately brings about the death of several human beings. That this was going to happen was made abundantly clear to the audience just before the mid-point intermission, and at intermission I went seething up the aisle toward a friend of mine I noticed in the audience.

In tones of deep shock, I said to him, "They're breaking First Law! They're breaking First Law!"

And my friend answered, calmly, "So why don't you strike them with lightning, Isaac!"

Somehow that restored my perspective and I watched the rest of the picture with something like calm and was even able to enjoy an arousal of curiosity.

Near the end of the picture when the spaceship was approaching Jupiter, several satellites were visible as small globes near the giant globe of the planet itself. I started counting the satellites at once, trying to figure out whether it was really possible to see them all in the sizes indicated from any one point in space.

Unfortunately, because they kept changing scenes, and because I could not remember the necessary data exactly enough or manipulate them without trigonometric tables, I could come to no conclusion.

So let's you and I work it out together now, if we can.

To begin with, Jupiter has twelve known satellites, of which four are giants with diameters in the thousands of miles, and the other eight are dwarfs with diameters of a hundred fifty miles or less.

Naturally, if we want to see a spectacular display, we would want to choose an observation post reasonably close to the four giants. If we do, then seven of the eight dwarfs are bound to be millions of miles away and would not be seen as anything more than starlike points of light at best.

Let's ignore the dwarfs then. There may be some interest in following a starlike object that shifts its position among the other stars, but that is not at all comparable to a satellite that shows a visible disk.

Concentrating on the four giant satellites, we will surely agree that we don't want to take up an observation post from which one or more of the satellites will spend much of its time in the direction of Jupiter. If that happens, we would be forced to watch it with Jupiter in the sky, and I defy anyone to pay much attention to any satellite when there is a close-up view of Jupiter in the field of vision.

For that reason, we would want our observation post in a position closer to Jupiter than are the orbits of any of the four giant satellites. Then we can watch all four of them with our back to Jupiter.

We could build a space station designed to circle Jupiter at close range and always watch from the side away from Jupiter, but why bother? There is a perfect natural station with just the properties we need. It is Jupiter's innermost satellite, a dwarf that is closer to the planet than any of the giants.

The four giant satellites of Jupiter were the first satellites to be discovered anywhere in the solar system (except for our own Moon, of course). Three of them were discovered on January 7, 1610, by Galileo, and he spotted the fourth on January 13.

Those remained the only four known satellites of Jupiter for nearly three hundred years. And then, on September 9, 1892, the American astronomer Edward Emerson Barnard detected a fifth one, much dimmer and therefore smaller than the giant four, and also considerably closer to Jupiter.

The discovery came as somewhat of a shock, for the astronomical world had grown very accustomed to thinking of Jupiter as having four satellites and no more. The shock was so great, apparently, that astronomers could not bear to give the newcomer a proper name of its own. They called it "Barnard's Satellite" after the discoverer, and also "Jupiter V" because it was the fifth of Jupiter's satellites to be discovered. In recent years, however, it has come to be called Amalthea, after the nymph (or goat) who served as wet-nurse for the infant Zeus (Jupiter).

Amalthea's exact diameter is uncertain (as is the diameter of every satellite in the solar system but the Moon itself). The usual figure given is 100 miles with a question mark after it. I have seen estimates as large as 150 miles.

For our purposes, fortunately, the exact size doesn't matter.

There is no direct evidence, but it seems reasonable to suppose that Amalthea revolves about Jupiter with one face turned eternally toward the planet. On half the surface of the satellite, Jupiter's mid-point is always visible. When standing on the very edge of that "sub-Jovian" side, the center of Jupiter is right on the horizon. The planet (as seen from Amalthea) is so huge, however, that one must go a considerable distance into the other hemisphere before *all* of Jupiter sinks below the horizon.

From roughly one-quarter of the surface of Amalthea, *all* of Jupiter is eternally below the horizon and the night sky can be contemplated in peace and quiet. For our purposes, since we want to study the satellites of Jupiter, we will take a position (in imagination) at the very center of this "contra-Jovian" side of Amalthea.

One object that will be visible in the contra-Jovian sky of Amalthea, every so often, will be the Sun. Amalthea revolves about Jupiter in 11 hours and 50 minutes. That is its period of rotation, too, with respect to the stars and (with a correction too small to worry about) with respect to the Sun as well. To an observer on Amalthea, the Sun will appear to make a complete circle of the sky in 11 hours and 50 minutes.

Since Amalthea revolves about Jupiter directly, or counterclockwise, the Sun will appear to rise in the east and set in the west, and there will be 5 hours and 55 minutes from sunrise to sunset.

With this statement, which I introduce only to assure you I am not unaware of the existence of the Sun, I will pass on to the matter of satellites exclusively for the remainder of the essay. The Sun has something to do with them, but what that something is, I will take up in the next chapter.

The four giant satellites, reading outward from Jupiter, are: Io, Europa, Ganymede, and Callisto. Sometimes they are called Jupiter I, Jupiter I, Jupiter III, and Jupiter IV respectively or, in abbreviated form, J-I, J-II, J-III, and J-IV.

Actually, for what we want, the abbreviations are very convenient. The names are irrelevant after all, and it is difficult to keep in mind which is nearer and which is farther if those names are all we go by. With the abbreviations, on the other hand, we can concentrate on the order of distances of the satellites in a very obvious way, and that's what we need to make the data in this article meaningful.

Using the same system, I can and, on occasion, will call Amalthea J-V. Generally, though, since it is to be our observation point and therefore a very special place, I will use its name.

So let's start with the basic statistics concerning the four giant satellites (see Table 12), with those for Amalthea also included for good measure. Of the data in Table 12, the least satisfactory are the values for the diameters. For instance, I have seen figures for Callisto as high as 3220 and as low as 2900. What I have given you is the consensus, as far as I can tell from the various sources in my library.

TABLE 12—THE FIVE INNER JOVIAN SATELLITES

Satellite	Name	Diameter (miles)	Distance from Jupiter's Center (miles)
J-V	Amalthea	100	113,000
J-I	Io	2,300	262,000
J-II	Europa	1,950	417,000
J-III	Ganymede	3,200	666,000
J-IV	Callisto	3,200	1,170,000

For comparison, the diameter of our own Moon is 2160 miles, so that we can say that J-I is a little wider than our Moon, J-II a little thinner, and J-III and J-IV considerably wider.

In terms of volume, the disparity in size between J-III and J-IV, on the one hand, and our Moon, on the other, is larger. Each of the two largest Jovian satellites is 3.3 times as voluminous as the Moon. However, they are apparently less dense than the Moon (perhaps there is more ice mixed with the rocks and less metal) so that they are not proportionately more massive.

Nevertheless, J-III is massive enough. It is not only twice as massive as the Moon; it is the most massive satellite in the solar system. For the record, here are the figures on mass for the seven giant satellites of the solar system (see Table 13). The table includes not only the four Jovian giants and our Moon (which we can call E-I), but Triton, which is Neptune's inner satellite and therefore N-I, and Titan, which I will call S-VI for reasons that will be made clear later.

If we are going to view the satellites, not from Jupiter's center (the point of reference for the figures on distance given in Table 12) but from the observation post on the contra-Jovian surface of Amalthea, then we have to take some complications into account.

TABLE 13—MASSES OF SATELLITES

Satellite	Name	Mass (Moon=1.0)
J-III	Ganymede	2.1
S-VI	Titan	1.9
N-I	Triton	1.9
J-IV	Callisto	1.3
E-I	Moon	1.0
J-I	Io	1.0
J-II	Europa	0.65

When any of the satellites, say J-I, is directly above Amalthea's contra-Jovian point, it and Amalthea form a straight line with Jupiter. J-I's distance from Amalthea is then equal to its distance from Jupiter's center minus the distance of Amalthea from Jupiter's center. This represents the minimum distance of J-I from Amalthea.

As J-I draws away from this overhead position, its distance from the observation point increases, and is considerably higher when it is on the horizon. The distance continues to increase as it sinks below the horizon until it reaches a point exactly on the opposite side of Jupiter from Amalthea. The entire width of Amalthea's orbit would have to be added to the distance between Amalthea and J-I.

Of course, from our vantage point on Amalthea's surface, we would only be able to follow the other satellites to the horizon. We will be faced with a minimum distance at zenith and a maximum distance at either horizon. Without troubling you with the details, I will present those distances in Table 14.

TABLE 14—DISTANCES OF THE JOVIAN SATELLITES FROM AMALTHEA

Satellite	Distance from Amalthea (miles)	
	At zenith	At horizon
J-I	149,000	236,000
J-II	304,000	403,000
J-III	553,000	659,000
J-IV	1,057,000	1,168,000

This change in distance from zenith to horizon is not something peculiar to Jupiter's satellites. It is true whenever the point of observation is not at the center of the orbit. The distance of the Moon from a given point on the *surface* of the Earth is greater when the Moon is at the horizon than when it is at the zenith. The average distance of the center of the Moon from a point on Earth's *surface* is 234,400 miles when the Moon is at zenith and 238,400 miles when it is at the horizon. This difference is very small because it is only the 4000-mile radius of the Earth that is involved. When the Moon is at the horizon, we must look at it across half the thickness of the Earth, which we need not do when it is at the zenith.

From a point on Amalthea's surface, however, we must look across a considerable part of the 113,000-mile radius of its orbit, which makes more of a difference.

In the case of our Moon, we are dealing with an orbit that is markedly elliptical so that it can be as close as

221,500 miles at one point in its orbit and as far as 252,700 at another point. Fortunately for myself and this discussion, the orbits of the five Jovian satellites we are discussing are all almost perfectly circular and ellipticity is a complication we don't have to face here.

Given the distance of each satellite from Amalthea, and the diameter of each satellite, it is possible to calculate the apparent size of each, as seen from our Amalthean viewpoint (see Table 15).

TABLE 15—APPARENT SIZE OF JOVIAN SATELLITES
AS SEEN FROM AMALTHEA

| Satellite | Diameter (minutes of arc) | |
	At zenith	At horizon
J-I	53	34
J-II	23	17
J-III	20	17
J-IV	10	9

If you want to compare this with something familiar, consider that the average apparent diameter of the Moon is 31' of arc. This means that J-I, for instance, is just slightly larger than the Moon when it rises, bloats out to a circle half again as wide as the Moon when it reaches zenith, and shrinks back to its original size when it sets. The other three satellites, being farther from Amalthea, do not show such large percentage differences in distance from horizon to zenith and therefore do not show such differences in apparent size either.

Notice that although J-III is considerably farther than J-II, it is also considerably larger. The two effects counterbalance as seen from Amalthea so that J-II and J-III appear indistinguishable one from the other in size, at least at the horizon. Or course, J-II, being closer, bloats just a little more when it reaches zenith. As for J-IV, it is smallest in appearance, and shows only one-third the apparent diameter of our Moon.

The sky of Amalthea puts on quite a display, then.

There are four satellites with visible disks, of which one is considerably larger than our moon.

But never mind size; what about brightness? Here several factors are involved. First there is the apparent surface area of each satellite, then the amount of light received by it from the Sun, and finally the fraction of received sunlight reflected by it (its albedo). In Table 16, I list each of these bits of data for each of the four satellites, using the value for our own Moon as basis for comparison.

TABLE 16—THE JOVIAN SATELLITES AND OUR MOON

Satellite	Apparent Area (Moon=1.0)		Sunlight Received	Albedo (Moon=1.0)
	Maximum	Minimum	(Moon=1.0)	
J-I	2.92	1.20	0.037	5
J-II	0.55	0.30	0.037	5.5
J-III	0.42	0.30	0.037	3
J-IV	0.10	0.084	0.037	0.4

If we consider the figures in Table 16, we see that J-I as seen from Amalthea is remarkable. At zenith it will possess an area up to three times that of our Moon. The intensity of sunlight it receives, however (as do the other Jovian satellites), is only ⅜₀ that received by the Moon. This is not surprising. The Moon, after all, is at an average distance of 93,000,000 miles from the Sun as compared to 483,000,000 miles for the Jovian satellites.

The Moon has no atmosphere and therefore no clouds— and it is atmospheric clouds that contribute most to light reflection. The Moon, therefore, showing bare rock, reflects only about 1/14 of the light it receives from the Sun, absorbing the rest.

The Moon's mark is bettered by J-I, J-II, and J-III. In fact, J-I reflects about ⅖ of the light it receives, which is every bit as good as the Earth can manage. This doesn't necessarily mean that these three satellites have an atmosphere and clouds like the Earth. It seems more likely that there are drifts of water-ice and ammonia-ice (or both) on the surfaces of the satellites, and that these drifts do the reflecting.

Callisto, for some reason, reflects only ⅟₃₀ of the light it receives and is therefore less than half as reflective as the Moon. Perhaps Callisto is composed of particularly dark rock. Or is it conceivable that astronomers have badly overestimated Callisto's diameter? (If it were smaller than astronomers think it is, it would have to reflect more light to account for its brightness.)

Anyway, we can now calculate the apparent brightness of each satellite (as compared with our Moon) by multiplying the area by the amount of sunlight received by the albedo. The results are given in Table 17.

TABLE 17—APPARENT BRIGHTNESS OF THE JOVIAN SATELLITES

| Satellite | Apparent Brightness (Moon=1.0) | |
	Maximum	Minimum
J-I	0.54	0.22
J-II	0.11	0.06
J-III	0.045	0.033
J-IV	0.0015	0.0012

As you see, not one of the Jovian satellites, as seen from Amalthea, can compare in apparent brightness with our Moon as seen from the Earth's surface. Even J-I, the closest to Amalthea and therefore the brightest, is never better than half as bright as the Moon; J-II is less than a seventh as bright; J-III less than a twentieth; and J-IV less than a six-hundredth.

And yet who says brightness is everything? Our own Moon is only ⅟₄₆₅,₀₀₀ as bright as the Sun, and if we consider beauty alone, it is all the better for that.

Perhaps the Jovian satellites as seen from Amalthea will be still more beautiful than our Moon, for being so softly illuminated. It will result, perhaps, in better contrast, so that craters and maria will be more clearly visible. If the satellites are partly ice-covered, patches of comparative brilliance will stand out against the darkness of bare rock. It will be all the more startling because on Amalthea there will be no air to soften or blur the sharpness of the view.

Callisto may be most beautiful of all, though it may require a field glass to see it at its best. It would be a darkling satellite, with its mysteriously low albedo. Perhaps it might look rather like a lump of coal, with its very occasional patches of highly-reflecting ice so interspersed by very dark rock that it will seem a cluster of diamonds in the sky, rather than a solid circle of light.

There are only two planets, other than Jupiter, that have real families of satellites, as opposed to merely one or two. These are Uranus with five and Saturn with ten. Uranus is a special problem (for reasons I mentioned briefly in Chapter 3), but let's tackle Saturn according to the system we have already used for Jupiter.

Although Saturn has only ten satellites to Jupiter's twelve, and only one giant as compared to Jupiter's four, it still puts on a better show in a way. Whereas no less than seven of Jupiter's twelve are so small and distant they can be ignored, only three of Saturn's need be neglected. From Saturn's innermost satellite, six other satellites can be seen with visible disks, and not four only.

Let's start by giving the basic statistics for the Saturnian satellites (see Table 18).

The Roman numerals are not as well established for Saturn as for Jupiter but I have seen them used from I through IX for the satellites from Mimas through Phoebe. Janus was discovered at the the very end of 1967 (see the preceding chapter) but I won't reorganize the numbering system because of that. Just as Jupiter's closest satellite is J-V, so I will let Saturn's closest satellite be S-X (even though it looks like a prudish way of writing "sex"). Besides, if we place our observation point on Janus (or S-X) it will be convenient to number the satellites in its sky as S-I, S-II, and so on.

If we assume that Janus presents one face, always, to Saturn and take up our position at the contra-Saturnian

TABLE 18—THE SATURNIAN SATELLITES

Satellite	Name	Diameter (miles)	Distance from Saturn's Center (miles)
S-X	Janus	300	98,000
S-I	Mimas	320	115,000
S-II	Enceladus	370	149,000
S-III	Tethys	800	183,000
S-IV	Dione	800	234,500
S-V	Rhea	1,100	328,000
S-VI	Titan	3,100	760,000
S-VII	Hyperion	250	922,000
S-VIII	Iapetus	750	2,213,000
S-IX	Phoebe	190	8,043,000

position, we will never see Saturn and its rings and we will be able to concentrate on the satellites.

We can work out the zenith and horizon distances of each satellite from Janus, as we did in connection with the Jovian system, and from that determine the apparent sizes of the Saturnian satellites (see Table 19).

As you see, the situation on Janus is most amazing. The outermost three satellites are only starlike points and are therefore omitted from the table. The other six satellites, which are included, are so closely spaced and increase in size so steadily as one goes outward that all appear, on the horizon, to be very much the same size. All have an apparent diameter about half that of our own Moon (S-I

TABLE 19—APPARENT SIZE OF SATURNIAN SATELLITES AS SEEN FROM JANUS

Satellite	Diameter (minutes of arc)	
	Zenith	Horizon
S-I	65	18
S-II	25	11
S-III	32	18
S-IV	20	15
S-V	17	13
S-VI	16	15

and S-III are a little larger, S-II and S-V are a little smaller, while S-IV and S-VI are just right).

This picture of sextuplet satellites is quite unique. Nothing like it can be seen from any other point in the solar system, not even from any other point in the Saturnian system.

Each of the six satellites bloats as it approaches the zenith, the effect being more extreme the closer the satellite. S-I expands from a diameter half that of the Moon at the horizon to twice that of the Moon at zenith. Its area (and therefore its brightness at any given phase) increases thirteen-fold, as it travels from horizon to zenith.

And the brightness of the Saturnian satellites? Here there is a difficulty that was not present in the case of the Jovian satellites, for there are no figures that I can find on the albedos of the Saturnian satellites. However, S-I and S-II are thought to be largely snow and SI-VI is known to have an atmosphere (the only satellite in the solar system *known* to have one).

We won't be too far out then if we decided to make the general albedo of the Saturnian satellites 0.5, or seven times that of the Moon. Working with that assumption and realizing that the Sun delivers only 0.011 as much light to the Saturnian satellites as to our own Moon, we can calculate the apparent brightnesses of the Saturnian satellites as seen from Janus (see Table 20).

Here we have a picture of a soft and delicate family of dim satellites, about as bright as Callisto (the dimmest of Jupiter's four giant satellites), as seen from Amalthea. All are only $\frac{1}{600}$ to $\frac{1}{200}$ as bright as the Moon. Only one of the Saturnians, S-I, manages to shoot up to the unusual mark of $\frac{1}{60}$ as bright as the Moon.

Does this give us all we need to know about the satellites of Jupiter and Saturn? Heavens, no!

So far I have painted only a static picture and left the most fascinating aspects of the situation untouched. Those four satellites of Jupiter as seen from Amalthea, and those six satellites of Saturn as seen from Janus, are moving relative to each other. Each moves at its own characteristic rate and the group forms an everchanging pattern.

What's more, the Sun moves across the sky, too (something I mentioned briefly near the beginning of the chapter) and that introduces interesting complications, too, such as phase-changes and eclipses.

I will try to work out the motion picture of the Jovian satellites, at least, in the next chapter.

TABLE 20–APPARENT BRIGHTNESS OF SATURNIAN SATELLITES

Satellite	Brightness (Moon=1.0)	
	Zenith	Horizon
S-I	0.0115	0.0032
S-II	0.0042	0.0020
S-III	0.0057	0.0032
S-IV	0.0035	0.0027
S-V	0.0030	0.0023
S-VI	0.0028	0.0027

7 The Dance of the Satellites

People always seem to be amazed over the fact that I am forever writing about spaceships but never get into an airplane; that I have my characters travel all over the galaxy while I myself drive to a neighboring state only with the greatest reluctance.

"You don't know what you're missing," they keep saying.

And they're right, I suppose, except that there's something also to be said for traveling in thought alone. It may not be quite as three-dimensional as the real thing, but it saves trouble.

For instance, about five years ago I drove the family to Niagara Falls, and the trip wasn't bad at that. We pulled into the neighborhood of the Falls, turned a corner, and there it was! I was fascinated; it was majestic; I was so pleased I had nerved myself to the 400-mile trip.

We got a nice pair of rooms in a motel in the very near neighborhood of the Falls and I lay down at last, so that I might sink into the rest I had so richly earned.

Or at least I tried to. My eyes closed, then opened, and a puzzled frown rested on my clear and ingenuous forehead. There was a dull roar that filled the room like a nearby train except that it was a train that neither approached nor receded, but remained where it was.

After a while, I identified the sound. With great indignation I realized that *they did not turn the Falls off at night*.

There you have it. An imagined Niagara may not be as impressive as the real thing, but it is quieter.

I shall never see the view from Amalthea which I described in the previous chapter—not even if a rocket ship to the Jovian system were available at this very moment with tickets for sale at a dollar a passenger and one of them reserved in my name. Still, I can *imagine* the view, and do so, gratis, in the peace and quiet of my attic.

While I'm at it then, I will now go on to consider the five inner satellites of Jupiter as moving bodies, rather than as static ones.

All revolve about Jupiter in fixed periods. The closer to Jupiter, the faster a satellite moves relative to the planet and the smaller its orbit about it; and, for both reasons, its period is shorter.

We can begin then by considering the period of revolution of each of the satellites and, in order to make those periods directly comparable, we can give them all in hours (see Table 21).

In traveling a circular path about Jupiter, each satellite sweeps over 360° (since every circle, whatever its size, can be divided into 360°). By dividing the period, in

TABLE 21–MOTION OF JOVIAN SATELLITES RELATIVE TO JUPITER

Satellite	Name	Period (hours)	Motion per Hour (degrees of arc)
J-V	Amalthea	11.83	+30.43
J-I	Io	42.45	+ 8.50
J-II	Europa	86.22	+ 4.18
J-III	Ganymede	171.7	+ 2.10
J-IV	Callisto	400.5	+ 0.90

hours, into 360°, we find how many degrees of arc each satellite traverses in one hour. That figure is given in the last column of Table 21.

The degrees of arc traversed in an hour are given as positive figures, because the satellites are moving in direct, or counterclockwise, rotation (see Chapter 3).

Suppose we imagine ourselves somewhere above Jupiter's equator, well above its unimaginably stormy atmosphere, so that we can observe the satellites in comfort. Suppose, also, that we are motionless with respect to Jupiter's center, which means we do *not* partake in Jupiter's rotation. All the satellites would then be seen to rise in the west, travel across the sky, pass overhead, then sink in the east.

But suppose we ourselves, while observing, were to circle Jupiter in the same counterclockwise direction. Even if we did so slowly, we would partially overtake the satellites which would seem to us, then, to move more slowly. They would still rise in the west and set in the east, but would take longer between rising and setting.

If we speeded our own motion more and more, the satellites would seem to move more and more slowly, as observed by ourselves from our moving vantage point. Finally, if we circled Jupiter at a speed that swept out +0.90° per hour, we would stay even with Callisto. The other satellites would continue to rise in the west and set in the east, but Callisto would seem to remain motionless in the sky as we matched it step for step. (Of course, we could still tell that Callisto was moving by comparing its position night after night with the neighboring fixed stars, but in this essay we are completely ignoring the motion of the satellites relative to the stars. Our view is confined entirely to the Jovian system itself.)

If we continued to hasten our motion, we would more than match Callisto—we would outrace it and it would seem to fall behind. It would rise in the east and set in the west and would seem thus to travel in a direction opposite to that of its real motion relative to Jupiter. (No mystery! If two trains were racing in the same direction on parallel tracks, passengers on the faster train would see the slower train seem to move backward as it is overtaken, even though it is really moving forward.)

As our own personal speed increased, we would next overtake Ganymede and force that into apparent east-to-west motion, then Europa, then Io, and so on.

But let's not set ourselves an arbitrary motion. Let us place ourselves on Amalthea, J-V, where we were in the preceding chapter. That will give us a fixed speed of +30.43° per hour. (If you're interested in a more easily visualized figure, Amalthea's speed about Jupiter is something like 16.7 miles per second relative to Jupiter's center, and this is nearly equal to the 18.5 miles-per-second figure of Earth's motion about the Sun. Compare this with Callisto, which moves at a speed of only 5.1 miles per second relative to Jupiter's center.)

Amalthea sweeps out many more degrees in a given time than any other of Jupiter's satellites, of course, and so it handily overtakes them all. From our vantage point on the contra-Jovian side of Amalthea, all the giant satellites would rise in the east and set in the west. (I am assuming, here, that Amalthea turns one face eternally toward Jupiter.)

We can easily determine the rapidity of this motion relative to Amalthea by subtracting the degrees-per-hour motion of Amalthea relative to Jupiter from the corresponding value for each of the other four satellites. Thus, if Amalthea moves +30.43° per hour relative to Jupiter and Callisto moves +0.90° per hour relative to Jupiter, then Callisto moves (+0.90) — (+30.43), or —29.53° per hour relative to Amalthea. The minus sign here would indicate that Callisto, as seen from Amalthea, was traveling east to west.

The motions of the various satellites as seen from Amalthea are given in Table 22. Having presented the names of the satellites in Table 21, I will henceforward (as I did in the previous chapter) continue to refer to them only by the Roman numeral identification.

As you see, the satellites all move quite rapidly from east to west, something which reflects Amalthea's very rapid motion from west to east. The closer a satellite to

Amalthea, the faster that satellite moves and the more effectively it tends to chase after Amalthea. None of the satellites does this very well, of course, but J-I, the closest to Amalthea, manages to fall behind only 22° per hour, whereas J-VI, the most distant, falls behind 29½° per hour.

TABLE 22—MOTION OF JOVIAN SATELLITES
RELATIVE TO AMALTHEA

Satellite	Motion per Hour (degrees of arc)
J-I	—21.93
J-II	—26.25
J-III	—28.33
J-IV	—29.53

It would seem then that the more distant a satellite from Amalthea, and the more slowly it turns about Jupiter, the more quickly it moves in Amalthea's sky. This may sound paradoxical, but all we are saying is that the more distant and slow a satellite, the more rapidly it is overtaken by Amalthea.

Let's put this into more familiar terms. From the motion in degrees per hour, it is not hard to calculate how many hours it would take to traverse 360°. This would move the satellite through a complete circle and give its period, in hours, relative to Amalthea (see Table 23).

TABLE 23—PERIOD OF JOVIAN SATELLITES
RELATIVE TO AMALTHEA

Satellite	Period of Revolution (hours)
J-I	16.4
J-II	13.7
J-III	12.7
J-IV	12.2

The period of revolution gives us the time lapse from satellite-rise to satellite-rise. (Actually, there's a small complication here in that Amalthea is somewhat over a hundred thousand miles removed from the center about which the satellites orbit. This means that a given satellite is 226,000 miles farther from Amalthea at some parts of its orbit than at other parts. Its motion as seen from Amalthea is not strictly uniform and the time from satellite-rise to satellite-set is not quite equal to the time from satellite-set to satellite-rise again. We will dispose of these complications by ignoring them.)

It so happens that Amalthea and the four giants all revolve in orbits that are nearly exactly in Jupiter's equatorial plane. Callisto's orbit is the most tilted but even so is only about ¼° off. (Compare this with our own Moon, which has an orbit tilted about 18° to the plane of the Earth's equator.)

This means that every time one satellite passes another in Amalthea's sky, there is an eclipse. (Actually, it is possible for Ganymede and Callisto to just miss one another in passing, sometimes, but even for them, a partial eclipse would be the rule.)

In order to determine the frequency of eclipses then, it is only necessary to calculate how long it would take one satellite to overtake another. It would always be the more distant satellite that would do the overtaking for it is the more distant that has the more rapid east-to-west motion. The more distant satellite would approach the nearer (and usually larger-in-appearance) from the east, slip behind it, and emerge at its west.

There would be six types of two-satellite eclipses altogether. The period between such eclipses, and the maximum time each would take from initial contact to final break-free, are given in Table 24.

I leave it to you to calculate (if you wish) the time between the various possible three-satellite eclipses: I-II-IV, I-III, and II-III-IV.

You might wonder why I've left out I-II-III, but apparently the three innermost giant satellites, J-I, J-II, and

J-III, do not move completely independently in their orits but maintain a certain fixed relationship that precludes their ever being in a straight line on the same side of Jupiter (although they can be in a straight line with two on one side of Jupiter and one on the other).

TABLE 24–SATELLITE-ECLIPSES AS SEEN FROM AMALTHEA

Eclipse	Time between Eclipses (hours)	Maximum Duration of Eclipse (minutes)
J-IV/J-III	300	25.0
J-III/J-II	173	20.7
J-IV/J-II	110	10.1
J-II /JI	83	17.5
J-III/J-I	56	11.7
J-IV/J-I	47	8.2

This eliminates the I-II-III eclipse and also (more's the pity) what would have been a most remarkable coming together of all four satellites in Amalthea's sky.

But now, what about the Sun?

Amalthea rotates about Jupiter in 11.83 hours, with (we assume) one side always facing Jupiter. That means it also rotates on its axis, relative to the Sun, in 11.83 hours. (The correction that needs to be applied as a result of Jupiter's motion in its orbit about the Sun in the course of that 11.83-hour period is so small it can be ignored.) As seen from Amalthea's contra-Jovian point then, the Sun rises in the east at 11.83-hour intervals, so that there is a 5.92-hour day and a 5.92-hour night. (The changing distance of Amalthea from the Sun in the course of the satellite's revolution about Jupiter is small enough to be neglected.)

To a viewer on Amalthea, the Sun would be just 6′ of arc in diameter, a bit less than one-fifth its diameter as seen from the Earth. This means that the Sun would be just visible as a distinct globe, rather like a glowing pea in the sky.

Its apparent area, as seen from Amalthea, would be only ½₇ of its apparent area as seen from Earth and it would therefore be only ½₇ as bright. The fall-off in brightness would be entirely due to the smaller apparent area of the Sun and not in the least due to the lesser brightness of the Sun itself, area for area. The pea-sized Sun of Amalthea would be just as bright as a similar-sized section of our own Sun would appear to be if the rest of it were blocked off. I would suggest then that our Amalthea-based viewer would not find it comfortable to stare at the Sun, shrunken though it might appear.

Amalthea's shrunken Sun would still be the incomparably brightest object in its sky.

To show that, let's abandon Amalthea's contra-Jovian point for a while and move to the other hemisphere, where we can see Jupiter. If the Sun were on the other side of the satellite, it would be shining over Amalthea's shoulder, so to speak, and lighting up the entire visible face of Jupiter. We would be seeing "full-Jupiter."

Jupiter would then be a shining glory, 43° across, or nearly one-quarter the full width of the sky. Its brightness would be −20.2 magnitude, or 1100 times as bright as our full Moon seen from Earth's surface.

The Sun, however, for all its shrunkenness, has a magnitude of −23.1 and would be 14 times as bright as Jupiter at its brightest. What's more, when the Sun and Jupiter are both visible from Amalthea's surface, Jupiter must be in less than its half-phase, so that it is considerably dimmer than it is at the full and is then even less able to compete with the diamond-hard brilliance of the tiny Sun.

Now let's get back to our contra-Jovian point, where Jupiter is never visible and where the Sun can only be compared to the satellites. The comparison is pathetic, in that case. Even Io, the brightest of the satellites (as seen from Amalthea) is never more than about half as bright as our full Moon. Amalthea's Sun is about 32,000 times as bright as Io at its brightest.

Furthermore, when the Sun is in Amalthea's contra-Jovian sky, all the satellites that are above the horizon are in the half-phase or less and are correspondingly dimmer. They are merely washed-out crescents.

As the Sun swoops across Amalthea's contra-Jovian sky from east to west, it does so faster than any of the satellites, faster even than Callisto. The Sun moves at a rate of —30.43° per hour (an exact reflection of Amalthea's motion about Jupiter, as you can see in Table 21).

The Sun therefore overtakes each of the four satellites, as shown in Table 25. If you compare Table 25 and Table 21, you will see that the length of times it takes the Sun to overtake a particular satellite is equal to the length of time it takes that satellite to circle the Sun once.

It is during the period from solar overtaking to solar overtaking that each satellite goes through its cycle of phases from new to full and back to new. The period is shortest for J-I, which also remains longest in Amalthea's sky. In the case of J-I, it can go from crescent at rising to nearly "half-Io" at setting. The other satellites change phase less spectacularly.

TABLE 25–THE SUN AND SATELLITES FROM
AMALTHEA

Satellite	Time between Solar Overtakings (hours)
J-I	42.45
J-II	86.22
J-III	171.7
J-IV	400.5

The Sun is farther from Amalthea than are any of the satellites, so that when the Sun overtakes a satellite, it can pass behind it and be eclipsed. This would happen every time if the Sun's apparent orbit were in the same plane as the orbits of the satellites.

However, all the satellites revolve in Jupiter's equatorial plane which is itself tipped 3° to the plane of Jupiter's orbit about the Sun. That means that the Sun's path across

Amalthea's sky interests the paths of the satellites in such a way that at places 90° from the points of intersection there will be a 3° gap between the Sun's path and those of the satellites. This is big enough to allow the Sun to miss the satellite completely so that there will be no eclipse.

However, every once in a while, the Sun will overtake a particular satellite so close to the point of intersection that the two will be only slightly separated (at the actual point of intersection they are not separated at all) and an eclipse will then take place.

How close to the point of intersection Sun and satellite must be depends on the apparent size of the Sun and satellite. For instance, the Sun has an apparent diameter of 6' of arc, and J-IV, at the horizon, one of 9' of arc. The distance, center to center, between the Sun and Callisto must be 7.5' if their edges are to appear to make contact. The distance, center to center, must be less than 2' of arc if Callisto is to eclipse the Sun entirely.

I am not enough of a celestial mechanic to make the appropriate calculations precisely, but by making some rough estimates (and I hope I'm not too badly off) I have worked out the data presented in Table 26.

TABLE 26—SATELLITE ECLIPSES OF THE SUN

Satellite	Number of Total Eclipses of Sun per Hundred Overtakings	Time between Total Eclipses (hours)
J-I	8	530
J-II	4	2,100
J-III	3⅓	5,200
J-IV	1	40,000

On the average, then, there will be one solar eclipse of some sort every 400 hours. The chances of seeing the Sun eclipsed by one satellite or another on any given Amalthean day is about 1 in 66. Such an eclipse is more common than on Earth and more easily viewed, too, because the satellites' shadows cut across all of small Amalthea, whereas the Moon's shadow narrows down to almost nothing by the time it reaches Earth's surface so that any given solar eclipse can be viewed from only a terribly restricted area.

The total eclipse doesn't last very long, because the apparent motion of the Sun in Amalthea's sky is so rapid. It endures longest when J-IV is the eclipsing body, since J-IV has the largest apparent motion of any of the satellites. It would take fully 5 minutes under the most favorable conditions, between the time when the last scrap of the eastern edge of the Sun disappeared behind Callisto and the first scrap of the western edge reappeared on the other side. For the other satellites, the eclipse never lasts more than 2 to 4 minutes.

A solar eclipse is not as spectacular on Amalthea as it is on Earth, in some ways. The Sun's corona is dimmer than it is here and it would be hidden by the satellite. (The impressiveness of a solar eclipse as seen from Earth rests largely in the extraordinary coincidence that the Moon and the Sun have almost equal apparent sizes so that the Moon just fits over the body of the Sun, allowing all the corona to be visible.)

Still, solar eclipses, as seen from Amalthea, will have their points. As the Sun approaches one of the satellites, the latter will show as a thin and shrinking crescent. If one or two other satellites are in the sky, farther removed from the Sun, they will be rather thicker crescents.

As the Sun disappears behind the one satellite, the other one or two would stand out brightly against a sky which now lacked the brightness of the small Sun.

What's more, there would be another phenomenon, considerably more interesting than that—a phenomenon which will take a bit of explaining. While the Sun is high in Amalthea's contra-Jovian sky, Jupiter is nearly full on the other side of Amalthea. (When the Sun is at zenith over the contra-Jovian point, Jupiter is entirely full.)

This means that the Jupiter-light shining on the satellites in the Amalthean sky is considerable. Ordinarily, a viewer at the contra-Jovian point on Amalthea would not be aware of Jupiter-light. During the Amalthean nighttime, the Sun would be on the other side of the satellite, near

Jupiter, and Jupiter would be a crescent. It would be comparatively dim, delivering little light to the satellites in the contra-Jovian sky. On the other hand, when the Sun was high in the sky and Jupiter was fat and bright on the other side of Amalthea, what light the planet delivered would be dimmed by comparison with the Sun's brilliance.

But what about the moment of solar eclipse, when Jupiter is bright indeed and the Sun is suddenly not there to compete? The satellite behind which the Sun is hidden, has its side-toward-Amalthea lit by Jupiter-light. To be sure, Jupiter-light is, at its very best, only $1/14$ as bright as sunlight (from the Amalthean viewpoint) but that is still rather impressive.

It would be particularly impressive under the circumstances. As the Sun passed behind a satellite, that satellite would be black by contrast and invsible against the black, airless sky except as a blot against the Sun. But then, when the Sun disappeared behind it altogether, the satellite would seem to flame out—suddenly visible in Jupiter-light.

Other satellites in the sky at the time would still be marked out as crescents by sunlight but the remainder of their Amalthea-facing surface would be lit by Jupiter-light and, with the Sun itself momentarily absent from the sky, that Jupiter-light would be more clearly visible.

Under favorable conditions, our own crescent Moon has the rest of its body very dimly lit by earthlight ("the old Moon in the new Moon's arms") but it must be remembered that Jupiter-light is considerably brighter than earthlight under comparable conditions. And in addition, the sunlit portions of the Jovian satellites would be less brilliant than the sunlit portions of our Moon.

So far I have described two types of eclipses, satellite-satellite and satellite-Sun. Let's consider a third type.

Imagine the Sun to be circling about Amalthea (as seen by an observer on Amalthea's surface) and let's imagine ourselves to be on the hemisphere facing Jupiter.

Each time the Sun makes its circle, it must pass behind Jupiter's swollen globe and would spend 1.4 hours behind it, too, so that there would be an eclipse for nearly one-quarter of the daylight period, during *every* daylight period. Nor would there be any Jupiter-light bathing Amalthea's surface, for during the period of the eclipse, Jupiter would be in its "new" phase, presenting only its dark side to Amalthea.

On Jupiter's contra-Jovian side, such an eclipse would never be directly seen, because Jupiter itself is never directly seen. When the Sun is in the contra-Jovian sky there is no chance for an eclipse (except the occasional momentary ones by the satellites). The contra-Jovian side of Amalthea gets the full 5.92 hours' worth of sunlight—one-third more than the Jupiter-side gets.

However, the eclipse of the Sun by Jupiter can be detected indirectly, for Jupiter's shadow falls on any of the satellites that may be on the night side of the planet.

From Amalthea's contra-Jovian side, it will be possible, at intervals, to see one satellite or another move into the Jovian shadow and blink out. Nor will it be lit by Jupiter-light, for, of course, it will be facing Jupiter's night side and the huge planet will be in the "new-Jupiter" phase.

I suppose an observer on Amalthea's contra-Jovian side, from his knowledge of the Sun and the satellites, which he could see, and from his observations of the manner in which each satellite would be eclipsed on occasion when the Sun was not in the sky, would be able to deduce the existence of Jupiter and get an idea of its size just from those eclipses, even if he never traveled to Amalthea's other side to see for himself.

8 The Planetary Eccentric

Some months ago, my wife and I saw *Plaza Suite*, a trilogy of three *very* funny one-act plays. We enjoyed them hugely in themselves, and we enjoyed them the more because the two leads were such favorites of ours and so good.

There are very few people who are not just a little star-struck, of course, and my wife has a feeling hidden deep within her that she can use me as a tool with which to meet those few certain stars that strike her. After all, she reasons, if one of them is a science-fiction fan, and if I walk boldly up and introduce myself . . .

The one catch is that I refuse to lend myself to this. After all, what if they are *not* science-fiction fans? Think of the humiliation!

So I left the theater, holding her elbow firmly and ignoring her plaintive plea that I make my way backstage and announce my magic name. We went to the restaurant next door instead for a cup of coffee while she glared daggers at me. And then the waitress told us in great excitement that the entire company of *Plaza Suite* was in that same restaurant.

Can I fight Fate? I sighed and removed my napkin. "All right," I said, resignedly, "I'll chance it."

I walked over with a charming smile, handed over my program booklet for autographs and casually introduced myself, pronouncing each syllable of my name with the greatest care. Needless to say, I bombed completely. There was no sign that any one of the crowd at the table had heard of me but as I drew back in chagrin, someone suddenly rose and dashed forward, shouting, "Gertie!"

My wife looked startled, then cried out, "Nate!" and, behold, they were grabbing at each other excitedly. Nate was her first cousin whom she hadn't seen in many years and he was right there at the table. She didn't need me at all. She had a great time and I stood to one side, shifting from foot to foot, and experiencing, but not enjoying, my unaccustomed role as husband-of-the-celebrity.

But we should all learn from our misfortunes and humiliations and to me it was just another case of getting practice in shifting viewpoints. The universe does *not* revolve about me. (I forget how many separate times I've very temporarily learned that.)

And it doesn't revolve about us, collectively, either, even though you'd never guess it from a casual glance at our astronomy texts.

For instance, any text will tell you how long it takes Jupiter to go around the Sun—a little under 12 years. And do they say: a little under 12 *Earth*-years? No, they don't.

As it happens, Jupiter goes around the Sun in one year by definition: one *Jupiter*-year. Of course, that sort of thing gives us very little information. Every planet goes around its primary in exactly one planet-year and that tells us nothing. We use the Earth-year as an arbitrary standard (convenient for ourselves) in order to compare periods readily.

But I wish, just once in a while, to see the matter explained straightforwardly, instead of having it all assumed so casually as to give the beginner the notion that what we call a "year" is an absolute length of time of cosmic significance instead of being a mere accident of astronomy.

Here's something else. If Jupiter's period of revolution is 11.862 Earth-years long, does it not follow that it is 4332 days long?

Yes, in a way, but it is 4332 *Earth*-days long. And in this case, our self-centeredness *does* deprive us of information. To give the period in Earth-days gives us nothing new as opposed to giving it in Earth-years. Why don't we instead give the period of Jupiter's revolution in *Jupiter*-days? That would give us something of astronomic value. It would tell us the number of times the planet rotates on its axis while it makes one revolution about the Sun.

I have never seen a table that gives the number of planet-days in a planet-year for each of the planets of the solar system (which doesn't mean such a table doesn't exist somewhere, of course), so I'll prepare one (see Table 27), omitting Mercury and Venus as special cases with unusual rotations (see Chapters 2 and 3).

As you see from the table, there are 10,560 Jupiter-days in a Jupiter-year, as compared with only 4332 Earth-days. You may deduce from this, and quite rightly, that the individual Jupiter-day is less than half as long as the individual Earth-day.

TABLE 27

Planet	Period of Revolution in Earth-years	in planet-days
Earth	1.000	365
Neptune	1.881	670
Mars	11.862	10,560
Jupiter	29.458	25,180
Saturn	84.018	68,130
Uranus	164.78	91,500
Pluto	248.4	14,192

The thing that catches the eye at once, though, in Table 27 is that the number of planet-days in Pluto's planet-year is extraordinarily low. Pluto has the longest period of revolution by far, yet has little more than half the number of planet-days in its year that Saturn has, and Saturn has a period only an eighth that of Pluto.

There's no mystery to this and I won't pretend to keep you in suspense. Pluto simply has an unusually long rotational period. This is surprising, too, considering that it is enormously far from the Sun and cannot, therefore, have been affected by the slowing effects of solar tides as Mercury and Venus were.

In Table 28, I list the rotational periods of the planets beyond Venus.

TABLE 28

Planet	Period of Rotation (Earth-days)
Earth	1.000
Mars	1.026
Jupiter	0.410
Saturn	0.426
Uranus	0.451
Neptune	0.66
Pluto	6.4

Clearly, something is wrong with Pluto. In fact, we can make that stronger. Everything is wrong with Pluto, and has been ever since it was discovered back in 1930.

As early as 1905, the American astronomer Percival Lowell had decided there must be some planet beyond Neptune (then the farthest known planet) to account for the fact that there were some small anomalies in Uranus's orbit. After all the gravitational effects of the other known planets* had been taken into account, something was left over.

Lowell calculated the orbit and mass such a planet would have to have to account for Uranus's departure

---

* Including Neptune, which was itself discovered as a result of Uranus's anomalous motions—but that is another story.

from its "proper" motion and figured out where it would have to be at that moment. He looked for it in that place, then all around that place, and didn't find it. By the time of his death in 1916, he still hadn't found it.

The fact that he didn't find it didn't necessarily mean it wasn't there. "Planet X," as Lowell called it, would have to be farther from us and more distant from the Sun, too, than any other planet was. Consequently, it would be dimmer than the other planets and an attempt to gather enough light to see it would also bring into view an extraordinary number of stars, against which it might well be lost.

Unless Planet X would be large enough to show a visible globe in the telescope, it would have to be detected by its motion against the background of stars, and here there was another catch. Since Planet X was so far from the Sun it would be moving more slowly in its orbit under the lash of the distant Sun's feeble gravity. Spotting its slow motion would not be easy.

In 1929, a young man, Clyde W. Tombaugh, began work at Lowell Observatory (which Lowell himself had had built) in Flagstaff, Arizona, and became involved in the search. He used a technique in which he photographed the same small part of the sky on two different days. Each picture would have from 50,000 to 400,000 stars in it. If all the stars were really stars, the two pictures should be completely identical. If one of the stars were really a planet, then that one spot of light representing it would have moved slightly during the intervals between photographs.

Tombaugh had the two plates projected alternately on a screen and adjusted them to have star images coincide. He then continued to project the images in quick alternation and began studying the picture painstakingly, area by area. Anything that was not a star would flick back and forth as the images alternated, and eventually that tiny flicker of shifting light would catch the eye. But it would have to flick back and forth only a small distance, for anything else that would be a non-star would move more quickly than Planet X.

On February 18, 1930, after almost a year of painstaking comparisons, Tombaugh found a flickering object in the constellation Gemini. For a month he followed that object and was then able to announce confidently that Planet X had been discovered. The announcement came on March 13, 1930, which would have been Lowell's seventy-fifth birthday if he had lived. Furthermore, the planet was named Pluto (a good name for a planet that swung so far from the light-giving Sun) and it was no accident that the first two letters of the name are the initials of Percival Lowell.

It was a great triumph for Lowell's memory, but the fact of the discovery was all the triumph there was. From that point on, everything went wrong.

In the first place, when its orbit was calculated, it turned out to be unexpectedly lopsided. The orbit departed further from perfect circularity and was more distinctly elliptical than was true for any other planet.

The ellipticity of an orbit (or of an abstract ellipse, for that matter) is given by a value called the "eccentricity." A circle has an eccentricity of 0 and an ellipse that is flattened into a straight line or elongated infinitely into a parabola has an eccentricity of 1. All ordinary ellipses have eccentricities between 0 and 1. For instance, the eccentricity of Earth's orbital ellipse is only 0.0168, which means that by casual inspection it couldn't be distinguished from a circle.

If a planet's orbit were a perfect circle, the Sun would be located at the exact center and the planet would be at the same distance from the Sun in every part of its orbit. When, however, a planetary orbit is elliptical, the Sun is located at one focus of the ellipse and this focus is displaced from the center of the ellipse. (It is this displacement which gives us the word "eccentric" which is from Greek words meaning "out of center.") The greater the eccentricity, the greater the displacement of the focus and the flatter, or more elongated, is the ellipse.*

This means that a planet moving around its orbit is closer to the Sun when it is on the side of the focus occupied by that body and farther when it is on the opposite side. There is therefore a point at one end of the major axis where the planet is at its closest to the Sun ("perihelion"), and another point at the other end of the major axis where it is at its farthest ("aphelion").

Thus, the Earth is 91,400,000 miles from the Sun at perihelion and 94,600,000 miles from it at aphelion. The difference in distance is 3,200,000 miles.

If the Earth's orbit were larger but retained the same eccentricity, the difference in extreme distances would be larger, too, but only in proportion. On the other hand, a greater eccentricity would make for a greater difference in extreme distances even if the average distance remained the same.

In other words, the difference in distance from the Sun, between perihelion and aphelion, is a measure of two things, the eccentricity of the planet's orbit, and its average distance from the Sun.

It so happens that Pluto is not only the farthest planet from the Sun, but it also has the most eccentric orbit. Usually these facts are presented in two separate columns, but I will give another column that combines the two facts to show you how enormous the effect is for Pluto (see Table 29).

You can see from the table that Pluto is nearly two billion miles closer to the Sun at some points in its orbit than at others. Since Pluto is less than a billion miles farther from the Sun, *on the average,* than Neptune is, you can further see that with Pluto swooping out nearly a billion miles farther than its average, and nearly a billion miles closer in than that same average, it is bound to

---

* There is a second focus, equally displaced, on the other side of the center along the "major axis"; that is, the longways diameter of the ellipse. In a circle, with an eccentricity of 0, the two foci fall exactly upon the center. All three are represented by a single point.

approach the Sun more closely than Neptune at some points in its orbit.

TABLE 29

Planet	Mean Distance (millions of miles)	Eccentricity of Orbit	Perihelion-Aphelion Difference in Distance (millions of miles)
Mercury	36	0.206	15
Venus	67	0.007	0.9
Earth	93	0.017	3.2
Mars	142	0.093	25
Jupiter	484	0.048	47
Saturn	887	0.056	103
Uranus	1790	0.047	168
Neptune	2800	0.009	48
Pluto	3680	0.249	1800

And so it does! At the present moment Pluto is moving toward its perihelion when it will be only 2,766,000,000 miles from the Sun. That point will be reached in 1989, and it will then be something like 35,000,000 miles closer to the Sun than Neptune will be. In fact, it will be closer to the Sun than Neptune will be through the entire range of years from 1979 to 1998. In that period, Neptune will be, temporarily, the most distant planet from the Sun.

Lowell had allowed for a certain eccentricity of the orbit of Planet X in order to make his figures come out right, but he didn't count on *that* much eccentricity.

When Pluto crosses Neptune's orbit in 1979 on its way inward, or in 1998 on its way outward, might it not collide with Neptune? Or if Neptune is in another part of its orbit then, might there not come a moment when the two planets approach the crossing point at the same time and have a catastrophic encounter?

From the usual picture of the solar system, we might think so, for it shows the orbit of Pluto making a shallow short cut across the orbit of Neptune in the neighborhood of Pluto's perihelion. However, the usual solar-system diagram is a two-dimensional projection of a three-dimensional phenomenon.

If you could view the planetary orbits from the side, you would see they are slanted at different directions. This slanting is usually defined by the angle of the plane of a particular planetary orbit and the plane of the Earth's orbit ("inclination to the ecliptic").

When a planet circles in an orbit so tipped, it is a given number of miles above the ecliptic at one end (at a point 90° from the place of crossing) and the same number of miles below it at the other end. This number of miles depends on both the size of the angle and the size of the orbit. Again, Pluto has both the largest inclination and the largest orbit, so that the results are spectacular (see Table 30).

TABLE 30

Planet	Inclination (degrees)	Maximum Distance from the Ecliptic (millions of miles)
Mercury	7.0	5.3
Venus	3.4	4.0
Earth	0.0	0.0
Mars	1.9	3.3
Jupiter	1.3	11.5
Saturn	2.5	41.0
Uranus	0.8	26.2
Neptune	1.8	88.5
Pluto	17.1	1340

Pluto wanders enormous distances above and below the ecliptic. No other planetary body can even faintly compare with it in this respect.

If Neptune and Pluto crossed the ecliptic at the same point and if that point just happened to be where Pluto crossed Neptune's orbit in the usual two-dimensional projection, then, yes, they would eventually collide. But such a coincidence would be fantastically improbable and it didn't take place. The points at which Neptune and Pluto cross the ecliptic are well separated, the closer pair (there are four crossing points altogether for the two planets) being over a billion miles apart, and none of the four being near the two-dimensional crossing point. This means that when Pluto's orbit seems to be cutting across Neptune's orbit in a two-dimensional diagram, the two orbits are many millions of miles apart in the third dimension.

With Pluto following so enormously deviant an orbit, is it possible that it can produce the effects on Uranus's orbit that Lowell had calculated under the assumption that the planet had a much more respectable orbit? The key to the answer lies in the mass of the planet.

To produce the effects on Uranus from the goodly distance at which Lowell expected Planet X to be, the new planet would have to be something like 6⅔ times as massive as Earth.

This is not in itself unlikely. The mass of Jupiter is 318 times that of Earth and the corresponding figures for the other three giants as we move outward from the Sun are 95, 15, and 17. Even allowing for a decline in mass as we move outward, should not Planet X have a mass of at least 6⅔ times that of Earth?

If a planet as dense as Neptune had a mass 6⅔ times that of Earth, it would have to have a diameter of 22,000 miles. If Pluto had this diameter and if it reflected 54 per cent of the light it received from the Sun, as Neptune does, then, even at its aphelion distance, Pluto would have a magnitude of 10.3 compared with Neptune's 7.6.

That is dim enough, heaven knows, but when Pluto was discovered it was considerably fainter than this even though it was considerably closer than its aphelion point. Even at perihelion its magnitude would be 13.6 and at aphelion it would dim further to 15.9. In other words Pluto was about 1/170 as bright as it ought to be.

To explain that, we must suppose that either Pluto is considerably smaller than 22,000 miles in diameter, and therefore catches less sunlight than we have assumed, or that it reflects considerably less of the sunlight than it receives—or both.

A small diameter would make it look bad, for then Pluto would be less likely to be sufficiently massive. Low reflectivity would also be bad for that would mean little or no atmosphere (which is what does most of the reflecting) and would therefore indicate a small gravitational field, hence a small mass.

This was something that could be understood at the instant of discovery, but it was the result of indirect reasoning. Could astronomers actually *measure* the diameter of Pluto?

Unfortunately, even large telescopes did not seem to magnify Pluto sufficiently to make it appear as a distinct little sphere. It remained nothing but a point of light (a bad sign of smallness in itself) for a quarter of a century after its discovery. This did not deprive astronomers of all information. At Lowell Observatory (where Pluto had been discovered) Robert H. Hardie and M. Walker detected small regular fluctuations in brightness in 1955 and from this they argued that the planet rotated in 6 days 9 hours. In 1964, Hardie sharpened the figure to 6 days, 9 hours, 16 minutes, 54 seconds, or almost exactly 6.4 days (the figure I used in Table 28).

Then, at last, in 1950, the Dutch-American astronomer Gerard P. Kuiper managed to produce a disk by looking at Pluto through the 200-inch telescope. From the size of the disk and the distance of Pluto, Kuiper decided that Pluto had a diameter of 3600 miles, which made it not very much larger than Mercury.

So small a planet could not possibly have the mass required to produce the effect on Uranus's orbit that Lowell had predicted. If Pluto had the density of Earth (the densest planet in the solar system) it would have a mass of only about 0.1 that of Earth. In order to have the required 6⅔ times Earth's mass, it would have to be over 60 times as dense as the Earth or some 15 times as dense as platinum, which is flatly impossible.

Besides, the low-mass hypothesis is further upheld by the fact that Pluto would then have no atmosphere to speak of and it might reflect no more of the light than our Moon does (say only 6 per cent of the sunlight that falls on it). It is the combination of small size and low reflectivity that accounts for the unusual dimness of the planet.

What is needed is a direct mass-determination, but this cannot be made.* If Pluto had a satellite, such a determination could be made in a matter of weeks, but it doesn't—at least none that we can detect.

Failing that, can there be anything wrong with Kuiper's size determination? One theory that was advanced was that Pluto might be completely ice-covered and might reflect the Sun so sharply that what we really see is the Sun's image as though in a fuzzy mirror and not the planet itself at all. This would mean that the planet was considerably larger than the image and might have the necessary mass after all.

Then an unusual chance to settle the matter came up. The Canadian astronomer, Ian Halliday, pointed out that Pluto was going to pass very close to a faint star on the night of April 28, 1965, so close in fact that it might actually pass in front of it and obscure it. If so, astronomers in different observatories, knowing how quickly Pluto moves, could compare notes on the length of time during which the star was hidden and from that calculate the true diameter of Pluto in a way that did not depend upon reflection of sunlight at all.

Came the great day and the star never disappeared! A dozen different observatories in North America agreed that the star remained visible throughout the period during which Pluto was in its neighborhood. Pluto had missed the star altogether!

If one claculated Pluto's exact position, however, it turned out that the planet had passed very close to the star and in order to miss the star even so, Pluto's globe had to be quite small. The diameter, by those calculations,

---

* Actually, just about the time this article appeared, a figure for the mass of Pluto *was* worked out by three astronomers at the U. S. Naval Observatory. It was 0.18 times that of the Earth.

would have to be less than 4200 miles, and Kuiper's original figures seem essentially confirmed.

Astronomers must, however reluctantly, accept Pluto as a dim planet, because it is a little one, and therefore non-massive, and therefore unable to account for the anomalies in Uranus's motions.

It follows, then, that Pluto is *not* the Planet X that had its orbit calculated by Lowell. It is an entirely different body that was discovered by the coincidence that it happened to be located within several degrees of the place where Lowell's figures had told Tombaugh to look.

So we are left with two questions. First, where is Planet X? Surely something is causing the perturbations in Uranus's orbit. Shouldn't astronomers do a little recalculating and relooking? Something fairly sizable may be out there.

The other question is: How do we account for Pluto and its eccentricities? It is completely different from the other outer planets in its orbital extremism, its mass, diameter, and even in its period of rotation.

Can it be that Pluto is not really a planet at all, but that it had once been a satellite? It is a little larger than the other satellites of the solar system but not much more so in some cases. Neptune itself has a satellite, Triton, which is about 2600 miles in diameter. Pluto would be only 2.5 times as massive as Triton if both were of equal densities.

Pluto comes suspiciously close to Neptune's orbit. Could it once have been a satellite of Neptune? If so, it would have probably faced one side eternally to its primary (as our Moon does). In that case, Pluto's present rotational period of 6.4 days would have represented the period of its rotation about Neptune. In order to revolve about Neptune in that period it would have had to be at a distance of 240,000 miles from the planet—a most reasonable distance.

Triton, Neptune's actual satellite, is only 220,000 miles from Neptune and revolves about that planet in 5.9 days. Actually, if Triton and Pluto both revolved about Neptune, they would be uncomfortably close to each other.

But there is something funny about Triton. It revolves

about Nephune with an inclination of 160°—that is, in the retrograde direction (see Chapter 3).

Some have suggested, from this one fact, that Triton was not always a satellite of Neptune but had been captured at some date in the past. Can it possibly be that the dynamics of the capture of Triton was such that Pluto, which had previously been Neptune's satellite, had been cast into outer space and made into an independent planet? Is that why Pluto's orbit is so eccentric and so inclined and why it homes back toward Neptune's orbit at every revolution?

Recently, new evidence has arisen.

First, it so happens that when a satellite moves about a planet in a retrograde fashion, while the planet itself rotates on its axis directly, tidal effects cause the satellite to move slowly closer to the planet. (Our own Moon, which revolves about us *directly*, is being slowly forced farther away by tidal effects, in contrast.)

In 1966, Thomas B. McCord of the California Institute of Technology calculated Triton's future and has decided that Triton may (in theory) crash on Neptune some time between 10,000,000 and 1,000,000,000 years from now. Actually, before it can really approach Neptune's surface, gravitational forces may well break Triton into fragments. Our remote descendants would then see a second planet with rings, and since Triton is more massive than the material in Saturn's rings, Neptune would end with a far more spectacular set than Saturn has.

Working backward, of course, we can see that Triton must have been further and further away from Neptune as we move farther into the past. McCord speculates that Triton may have been so far away long ago as not to have been part of Neptune's system at all. Triton would then have been captured by Neptune long ago.

And when it was captured, did it knock out Pluto? There are, alas, objections to that pretty picture. Pluto does not really come very close to Neptune after all, if the solar system viewed three-dimensionally, as I explained earlier.

To be sure, further gravitational influences on Pluto

since the catastrophe may well have altered Pluto's orbit still more—but *that* much?

In 1964, astronomers at the U.S. Naval Weapons Laboratory in Dahlgren, Virginia, calculated Pluto's orbit backward for 120,000 years to see if various gravitational influences of other planets might have had it significantly closer to Neptune in the past. They did not!

But that still leaves Planet X. I think there is a good chance Planet X may be out there somewhere, perhaps at the other end of Pluto's eccentric orbit. If Planet X can be found and it's gravitational influences taken into account and Pluto's orbit calculated back for millions of years rather than merely thousands, why, who knows? It may turn out that the various puzzles of Pluto so frustrating now will fall into neat place, once one more planet is taken into account.

II
—AND BACK

^A PHYSICS

9 Just Right

I was walking along the street one day recently, making my way briskly toward some destination or other and, as is sometimes my wont, let myself sink deeply into thought.

Now, I don't know what expression you may wear when you are in a state of absorbed reverie, but I am told that my own face, under such conditions, wrinkles into an expression of unbelievable savagery. I find this incredible, since I am notorious for my sunny disposition and for my carefree, happy-go-lucky nature, but I suppose there must have been some reason why my children (when they were younger) would rush screaming away from the dinner table whenever a knotty point in my writing occurred to me and required thoughtful resolution.

On this particular occasion, as I was walking along absorbed in thought, a perfect stranger, walking in the opposite direction on business of his own, said to me as we neared, "Smile!"

I stopped short, smiled, and said, "Why?"

And he said, with a smile of his own, "Because nothing, but *nothing*, can be as bad as all that."

We separated and I did my best to continue thinking

133

and smiling, too; but I suspect that little by little the smile faded and the savage look returned. . . .

I did, though, out of curiosity, take special note of what it was that I was thinking of then, in order that it might (if possible) become the subject of a science essay.

It turned out I was thinking of a new TV program called "Land of the Giants," in which a party of human beings is trapped on a world which is just like Earth except that everything is of giant size. To be specific (I checked with the producers of the program) everything on the giant world is scaled at a ratio of 12 to 1 compared to analogous objects on Earth.

This carries to an extreme a well-known type of plot in what we might call "infantile science fiction." By that phase, I refer to the kind of science fiction produced by men who are undoubtedly kind to their mothers and who are estimable members of society, but who, as far as their understanding of science is concerned, are drooling babies.

Back in the bad old days of magazine science fiction, there were innumerable stories about giant insects, for instance. The reasoning was that since a flea could jump many times its own length, and pull many times its own weight, a flea that was of human size could jump half a mile with two tons of stuff on its back. And, of course, he would be far more dangerous than a tyrannosaurus. Needless to say, this is thorough hogwash, and you will find such nonsense nowhere in the s.f. magazines of today.

Movies and television, however (with some notable and honorable exceptions such as "Star Trek") are still in the infantile stage as far as science fiction is concerned. Their idea of excitement is to give us giant apes, giant spiders, giant lizards, giant crabs, giant women, giant amoebas, giant anything.

And none of it would work for a moment because of something called the square-cube law, which was first explained by Galileo three and a half centuries ago.

To show what the square-cube law means in the simplest possible way, let's start with a cube, each edge of which is n inches long.

The volume of this cube is n x n x n, or n^3. This means that a cube with a 1-inch edge has a volume of 1 cubic inch; one with a 2-inch edge has a volume of 8 cubic inches; and one with a 3-inch edge has a volume of 27 cubic inches. Or, to put it another way, you can take one 3-inch-edge cube and saw it into twenty-seven 1-inch-edge cubes. Try it and see.

What about the surface of the cube, however?

The surface consists of six square faces (which is why dice have their faces numbered from · to :::). If the edge of such a cube is n inches long, then each face has an area of n x n, or n^2, and all six faces have an area of $6n^2$. This means that a 1-inch-edge cube has a surface of 6 square inches; a 2-inch-edge cube has a surface of 24 square inches; a 3-inch-edge cube has a surface of 54 square inches, and so on.

Since an n-inch-edge cube has a surface area of $6n^2$ and a volume of n^3, this means that the surface of the cube increases as the second power (or square) of the length of the edge, while the volume of the cube increases as the third power (or cube) of the length of the edge. If you double the length of an edge of a cube, you increase its surface by 4 times (2^2), but its volume by 8 times (2^3). Similarly, if you triple the length of an edge of a cube, you increase its surface by 9 times, but its volume by 27 times.

The volume increases much faster than the surface and just to pound away at that some more, here is Table 31 showing it:

TABLE 31

Edge (length)	Surface (area)	Volume	Volume/Surface
1	6	1	⅙
3	54	27	½
6	216	216	1
10	600	1,000	1⅔
25	3,750	15,625	4⅙

The larger a cube is, then, the more volume it has for every square inch of surface. The larger it is, the larger the percentage of its substance is on the inside, so to speak.

You can show exactly the same thing to be true of any other geometrical solid—a tetrahedron, a sphere, an ellipsoid, and so on. It is even true of any irregular solid, provided (and this is an important provision) that the solid retains its exact proportions as it grows larger.

We can state the square-cube law as follows then: As any three-dimensional object increases in size without any change in proportion, the surface will increase as the square of the linear measurement and the volume will increase as the cube of the linear measurement.

This has an important relationship to structural engineering, both in animate and inanimate objects, for some properties of such objects depend on the volume and some on the surface. Since the volume-dependent properties increase faster with size than the surface-dependent properties do, there are many times when size makes a considerable difference.

The simplest example is mass and support. The mass of any object (or its weight, if it remains in a fixed point on the earth's surface) of fixed shape and density depends on its volume. Its support depends on the area of the part that makes contact with the ground.

For instance, let's imagine a cube of substance that has a density of 1 pound per cubic inch. A 1-inch-edge cube of this substance, resting on one of its faces, weighs 1 pound and rests on a face that is 1 square inch in area. The pressure on that supporting face is 1 pound per square inch.

A 10-inch-edge cube of this substance weighs 1000 pounds and rests on a face that is 100 square inches in area. The pressure on that supporting face is 10 pounds per square inch.

As the cube continues to increase in size, the pressure on the supporting face continues to increase as well. Eventually, the pressure on that supporting face becomes so large that the chemical bonds between the atoms and molecules of that substance give way. The cube begins to flatten under the pull of gravity.

The greater the tensile strength of a substance, the larger it can grow before this crucial point is reached, but for all substances this crucial point will be reached eventually. In a given gravitational field, there is a maximum size for any cube of any given substance.

This is true even if there is no outside gravitational field, for as a cube increases in size its own gravitational field increases and forces the cube to "flatten out" or, rather, to assume a shape of minimum energy content. This turns out to be an approximation of a sphere (or, more accurately, an ellipsoid of revolution).

What holds for a cube holds for all other solids—including a human being.

Consider someone who weighs 175 pounds and has a pair of feet with soles that have a total surface area of 50 square inches. When he is standing, each square inch of the sole of his feet is supporting 3.5 pounds. (This is a simplification. The soles aren't flat and weight is not evenly distributed on them, but that doesn't alter the principle.)

Now suppose that this human being is suddenly expanded twelve-fold in every dimension (as in "Land of the Giants") with all his parts remaining in their original proportions. Instead of being 5 feet 10 inches tall, he is now 70 feet tall.

The giant's weight is now 175 x 12 x 12 x 12, or 151.2 *tons* (as much as the largest whale in existence). The surface area of his feet, however, is only 50 x 12 x 12, or 7200 square inches. When he is standing, each square inch of his soles must support 42 pounds, twelve times as much as before.

This holds for other supportive machinery. Each square inch of thighbone cross section must support twelve times the weight it ordinarily does; each square inch of muscle cross section must exert twelve times the pull it ordinarily does if such a giant is to stand up from a sitting position and so on.

To see what would happen to such a giant, suppose you

placed 42 pounds (twelve times the normal) on each square inch of your soles. To do this, you would have to have a weight of one ton evenly distributed over your body. That would drop you to the floor and crush you to death.

Well, the giants of "Land of the Giants" would drop to the floor under their own weight and be crushed to death. Oh, not necessarily, of course. If a competent science-fiction writer were doing the series, he'd hint that the giants' thighbones were made of chrome steel, that the planet's gravity was somewhat weaker, that their muscles worked on some different principle from ours. But none of that is done, of course. I said, a *competent* science-fiction writer.

If you think I'm being too pessimistic about the crushing to death, it actually happens. Sometimes a whale (one that is considerably smaller than our mythical human giants) gets itself stranded on a beach. It then proceeds to die because it is literally crushed by its own weight.

(In water, the whale has no problem. In water, the supportive influence is not the rigidity of bone, but the buoyancy of the liquid medium. The amount of buoyancy depends upon the *volume* of the organism. This means that weight and buoyancy *both* increase as the cube of the linear dimension, so that size is not important as far as support is concerned. A huge whale maneuvers through water with no more difficulty than a tiny minnow does— at least as far as support goes.)

Of course, even if we restrict ourselves to land mammals, the fact remains that there are both dwarfs and giants built on the same general structural plan of a trunk and four legs. There are mammals as small as a shrew (less than a tenth of an ounce) and as large as the extinct baluchitherium, which may have weighed as much as twenty tons.

How can such variations in size be squared with the square-cube law? Well, remember the condition—that there be no change in proportion or (let us now add) in structural properties.

Perhaps the most remarkable case of neglecting that condition involved the Canadian-American astronomer Simon Newcomb, who in the first few years of the twentieth century wrote a series of eloquent articles in which he tried to debunk the gathering excitement over the possibility of constructing heavier-than-air flying machines.

Newcomb painstakingly pointed out (much as I am doing in this article) the existence of the square-cube law. He explained that the weight of an airplane depended on its volume, while its support depended on the area of flat surface it could present to the air. As it increased in size, the weight increased more rapidly than the surface area and each square inch of the wing surface would be required to support a larger and larger weight.

By the time an airplane was large enough to hold a man, said Newcomb, it would be too heavy to be supported by its wings. It seemed as simple as grade-school arithmetic to him.

Came the Wright brothers, however, and it became apparent that a flying machine large enough to carry a man was possible. This did not stump Newcomb at all. He admitted a machine that large was possible after all, but by the time it was made large enough to hold two men, he said, it would be too heavy to be supported by its wings.

Newcomb died in 1909 and did not live to see the airplane come into its own in World War I.

Newcomb's mistake was a common one (alas) among scientists. Enamored with a relationship, he insisted on carrying it beyond the limits to which it was applicable. He assumed that as airplanes were made larger, its proportions would remain the same; that the materials of which it was made would stay the same, and so on.

But, as a matter of fact, airplane wings were continually being improved so as to offer more lift per square inch; engines were designed to give more thrust per unit weight; stronger, but less dense materials were designed for the body of the plane. In short, through improved engineering, support was increased much more rapidly than the square of the linear dimension, and weight was increased much less rapidly than the cube of the linear dimension.

Living organisms do the same thing, insofar as possible. The bee hummingbird weighs only 0.07 ounce, while the kori bustard of South Africa, it is suspected, may on occasion reach a maximum weight of perhaps 50 pounds. That is an eleven-thousand-fold spread in weight. Yet both hummingbird and bustard fly on feathered wings.

But there's a difference. The bustard wings are much longer and narrower in proportion to the body than the hummingbird wings are. The bones of the bustard are hollowed to the last bit so as to be made as light as possible at the expense of as little as possible of their strength. Again, these are changes that make it possible for weight to increase less rapidly than the cube; support, more rapidly than the square.

In the case of mammals, the spread from pygmy shrew to baluchitherium is 6,400,000 to 1 for the same general scheme of trunk on four supports. But those supports are not identically proportioned by any means.

If you look at the legs of a shrew, a mouse, a goat, a horse, and an elephant in that order, you will see that they grew thicker and thicker even in proportion to the overall body length. If you could see a picture of each animal drawn to the same height, this would be plain to you.

If a mouse were expanded to the size of an elephant, its legs would be spindly in comparison and would break under its weight like toothpicks. An elephant decreased to the size of a mouse, would have stubby legs that would be incredibly clumsy.

In other words, an animal's proportions—the shape of its legs and its wings, for example—are conditioned by its size, and would be all wrong if that size were changed without an appropriate change in proportions.

So do you really expect man-sized insects would be dangerous? Next time you see a housefly look at its legs. They are mere threads, but adequate to hold up the fly's weight. Increase the fly's size to that of a man and it couldn't move. Neither could a flea or a grasshopper or a beetle. No insect the size of a man that retained the proportions of ordinary insects, could (in the Earth's gravitational field) walk, fly, hop, or otherwise progress the smallest fraction of an inch.

It's not just support and locomotion that depend on size and proportion. There are numbers of other properties that are designed to be just right for a particular size.

For instance, the amount of heat produced by the chemical reactions within a body depends on the weight of reacting tissue in the body which, in turn, depends upon its volume. The rate at which such heat is lost depends upon the surface area of the body (roughly speaking). This means that the larger the animal, the more heat it retains, since the production rises faster than the loss.

In general, then, all other things being roughly equal, a small animal must have a faster metabolism than a large one if it is to replace the more-quickly-leaking energy. A shrew or a humming-bird must be constantly eating and will die of starvation in a matter of hours while a large animal can last for long periods.

It also follows that in arctic areas where the cold temperatures favor devices that lead to conservation of heat, large size is of particular value in this respect. The walruses and polar bears and musk oxen stay warm partly by being large. And, as a matter of fact, this may have been one of the factors that made large size useful in the age of the dinosaurs. The reptiles, with no special devices to maintain high temperature, could retain what heat they developed with greater efficiency the larger they were.

This means that the whole metabolic structure of an organism is related to its size.

Then, too, what about the absorption of oxygen? The weight of material requiring oxygen depends on the volume of the organism, but the rate at which oxygen is absorbed depends on the internal surface area of the lungs.

Simple baglike lungs suffice for small cold-blooded animals, but warm-blooded animals need more oxygen, and large warm-blooded animals need far more. If the human lungs were simple bags, they would offer about 2 square feet of surface for the air, and if that were all there was, we would suffocate almost at once. Our lungs are broken up into about 600,000,000 tiny little chambers, and the total surface of all those chambers is at least 600 square feet.

In the same way, the quantity of blood to be filtered depends on the weight and, hence, volume of the animal. The rate at which it can be filtered depends on the surface area available in the kidney. For that reason, the kidney is broken up into over a million little tubes and their total length in both kidneys comes to about 40 miles.

Therefore if a man's measurements were suddenly multiplied by 12 in every direction, with no other change at all, he would asphyxiate in minutes, for his lung surface would be increased 144 times whereas the quantity of body to be fed oxygen would be increased 1728 times. And if he survived that he would die of uremia in days, for he would have the surface of his kidney's filter tubes increased 144 times while the volume of blood to be filtered would be increased 1728 times.

No, a giant man, even if he had his thighbones thickened and his feet splayed out enormously to bear the weight of all his tonnage, would still have to have immensely more complicated lungs and kidneys, and for that matter immensely more involved networks of blood vessels and nerves.

And insects? Insects are ventilated by small tubes in the abdomen where ordinary diffusion is just about enough for the creature's needs. Expand an insect to the size of a man without utterly changing its respiratory system and it would asphyxiate at once. Indeed, nothing would be as utterly helpless, harmless, and *dead*, as that great science-fictional menace, the giant insect.

Consider man's crowning possession—his brain.

Man's brain is one of the largest in existence, a little over 3 pounds in weight, but not quite the largest. A large elephant may have a brain that weighs about 13 pounds and the largest whale brain could weigh 19 pounds. More important, however, is the pounds of body that must be coordinated by each pound of brain.

The body/brain mass-ratio is about 50 in man. That is, there are 50 pounds of tissue to be taken care of by each pound of brain. The corresponding figure for a large elephant is 1000; and for a giant whale, 10,000. (The largest dinosaurs had a body/brain mass ratio of 100,-000.)

Here, at least, we might seem on safe ground and can avoid the square-cube law. As the body increases in dimensions, the total weight varies as the cube of the linear dimension and so does the weight of the brain. The body/brain mass-ratio would remain 50 even in the 70-foot giants of "Land of the Giants."

But the brain cells that make up the crucial part of the brain—the gray matter, that is—are concentrated on the surface of the cerebrum. For the highest functions of the brain (from our own prejudiced standpoint), for thought and intelligence, that is, what counts is not the weight of the brain after all, but its surface area.

As intelligence increases, the surface area of the brain must increase faster than the square-cube law would allow and it can do this only by forming wrinkles or convolutions. The gray matter, as it dips in and out of those convolutions, is present in greater quantity than if it were stretched smoothly over the cerebral surface.

For that reason, the presence and number of convolutions is a way of estimating intelligence and the human brain is not only larger than almost any other creature's; it is also more convoluted.

If we expand a man to twelve times his dimensions in every direction and if the brain expands as well in every dimension, the brain will remain heavy enough in proportion but the surface will fall behind. Unless the brain becomes twelve times as convoluted, it won't maintain adequate control of the larger body. If a 12-times-larger giant simply expands his brain without change, he will be an utter idiot, despite a brain that will weigh about 2.8 *tons!*

In short, then, largeness isn't such a great thing. It complicates matters enormously in every respect and, after a certain point, the advantages to be gained from size (such as better heat retention, larger eyes and therefore more acute vision, larger brains and therefore greater intelligence) begin to be overbalanced by the disadvantages of ever-expanding complication.

I like to think, with my usual prohuman orientation, that the size of the human being is *just right!*

(Among the sea creatures, where support against gravity is no problem, the point at which the disadvantages of size begin to outweigh the advantages is higher up on the scale. On the whole, then, sea creatures tend to be larger than land creatures and the largest of all animals have lived in the sea rather than on land.

But what if we move in the other direction?

If we decrease a man's size in all directions, would not the surface area of his brain have less body to handle in proportion? If a 12-times-enlarged person, unchanged otherwise, becomes an idiot, would not a 12-times-diminished person, unchanged otherwise, be a super-genius?

Ah, but a 12-times-reduced man will have a brain weighing about 0.03 ounce. It will contain only 6,000,000 neurons rather than 10,000,000,000. And, no matter how convoluted the brain and how little body you have to take care of, 6,000,000 neurons can't be hooked up in a complicated-enough fashion to allow human intelligence.

In other words, the absolute weight of the brain also counts, and we have examples of that. In some of the smaller monkeys, the body/brain mass-ratio is only 17.5. If such a monkey were expanded to man size, his brain would weigh 8.5 pounds. And yet a small monkey is far less intelligent than a gorilla with a body/brain mass-ratio of 500. The monkey brain isn't as convoluted as the gorilla brain (let alone ours), but in addition, it just doesn't have enough cells.

No, our size is just right. Neither too large nor too small.

It may well be that you think you have me now. A couple of years ago, I was responsible for a novel that dealt with the drastic miniaturization of human beings to less than bacterial size. You may now be thinking that I did not then practice what I am now preaching.

If so, you are quite wrong. The drastic miniaturization of human beings (assuming such a thing were possible) involves a number of fascinating little physiological points, which I tried to take into account in the novel, and which I will explain in the next chapter.

10 The Incredible Shrinking People

In April, 1965, I was asked to prepare a novelization of the script of a movie that was then in production. The movie was *Fantastic Voyage* and it eventually received a couple of Oscars for special effects.

To put the plot in a nutshell, a submarine and its crew of five are miniaturized to microscopic size and injected into the blood stream of a dying man in order that they might perform a brain operation from within and save his life. They had exactly sixty minutes to do it in for at the end of the time the miniaturization effect would wear off. If they were still inside the man at that time, their re-expansion would, of course, kill him.

Naturally, there are all sorts of untoward events that delay the operation and, in the end, the crew get out of the man (whose life they save) with something like two seconds to spare.

I had never made a novel out of a movie script before and I am all but incapable of resisting the chance of tackling something new—so I eventually let them talk me into doing it.

I read the script and said, "I will have to change the ending, if I do the novel."

They were alarmed at once. "Why?"

"Well," I said, "at the end, the ship and the villain are ingested by a white corpuscle and the other four get out. Right? But the ship and the villain stay inside. I'll have to get them out, too."

That puzzled them. "Why?"

"Because the ship and the villain will expand if they stay inside the patient, and that will kill him."

They thought about that awhile and then they said, "But the white corpuscle *ate* them."

I said, "That doesn't matter; the atoms are still there and as long as they're still there, even if they are all separated and evenly scattered—"

Then I stopped, because I realized that they were staring at me blankly. I said, "Look, I'm going to change the ending. If you don't want me to change the ending, fine; I won't do the book. But if you want me to do the book, I will change the ending, and I don't want my ending changed back by Hollywood. Okay?"

So they said "Okay," and in the book I managed to work out a way of getting the white corpuscle, *with* the ingested ship and villain, out of the patient. Nor did Hollywood change it back. Indeed, Hollywood didn't change one word of my novel, I am glad to say.

In the motion picture, however, the ship was still left inside the patient.

This had its annoyances, too, for my book (I write quickly) came out six months before the movie (they work slowly) so that everyone thought the movie had been made from the book, rather than vice versa. People who saw the movie and didn't read the book therefore wrote me shocked letters about the ending, and I had to answer them patiently.

The whole business of "shrinking"—a well-known science-fictional motif—is based entirely on several indefensible assumptions, such as the assumption that the law of conservation of energy can be ignored and that atoms don't exist.

Let us suppose, for instance, that we have the kind of situation that was postulated in such well-known science-fiction stories as Ray Cummings' antiquated "The Girl in the Golden Atom" or in Henry Hasse's excellent "He Who Shrank," or in the well-done Richard Matheson movie *The Incredible Shrinking Man* (the title of which I borrowed with modification for this chapter).

In these and other stories the shrinkage is carried into the submicroscopic, but we shall be moderate about it and begin by supposing that a man is shrunk to exactly half his ordinary height.

In that case he is also shrunk to exactly one-eighth his ordinary volume (see the preceding chapter). There are three alternatives as to what can have happened to account for that volume shrinkage:

1. Perhaps the atoms making up his body are squeezed more closely together.

2. Perhaps seven-eighths of the atoms of his body, drawn proportionately from all parts, are discarded.

3. Perhaps the atoms themselves shrink.

The first possibility is reminiscent of the situation in gases. A volume of ordinary gas can, without too much difficulty, be compressed to one-eighth its volume by squeezing out most of the space between its atoms or molecules.

However, though the atoms and molecules in gases are widely spread apart and can easily be forced more closely together, those in liquids and solids are in virtual contact and can be pushed more closely together only very slightly even by enormous pressures. The pressure at the center of Jupiter might suffice to reduce the volume of a man considerably, but not nearly down to one-eighth normal. Before that could come to pass, the atomic structure itself would break down.

To be sure, once the atomic structure breaks down, shrinkage can continue to very small volumes, something that takes place in the interior of stars. Unfortunately, human beings would find difficulty in surviving such conditions, wouldn't they?

So throw out possibility one.

The second possibility seems much more plausible. You just go through the human body and retain one out of every eight molecules, keeping all the different types of molecules in proper proportion. (It would be like taking a Gallup poll of the body.)

This assumes, however, that the human body can get along on only one-eighth of its molecules. To be sure, a much smaller number of molecules than those we possess is compatible with life. Mice live; bacteria live.

But what about the brain? The normal brain has a mass of 3 pounds. Get rid of seven-eighths of all its molecules and what is left has a mass of 6 ounces. It doesn't matter how carefully you keep the brain-molecules present in appropriate proportions, a 6-ounce brain is not large enough to maintain intelligence at human level.

And if you're going to quarrel with that, then what about reducing the human being still further, as is done in every science-fiction story that deals with shrinkage?

Throw out possibility two.

That leaves only the third possibility, that the atoms themselves shrink. In that case, the shrinking people have their atoms neither forced more closely together nor discarded one after the other. To themselves, they remain as they always were. In terms of atoms, they are just as comfortably-spaced and just as numerically-complex as ever.

It is possibility three that I specifically stated I was using in the novel version of *Fantastic Voyage.* (The movie ignored the whole matter.)

But we're not out of the woods. Suppose the atoms themselves shrink. What happens to their mass? Here there are two possibilities:

1. Perhaps the mass remains unchanged.
2. Perhaps the mass shrinks in proportion to volume.

The first possibility would quickly produce unacceptable complications. It would mean that a 6-foot, 200-pound man would be reduced to a 3-foot, 200-pound man. Instead of being roughly as dense as water, he would become roughly as dense as steel, and he would become even denser as he shrank. By the time he was 2 feet high, he would be denser than platinum, and if he were reduced to microscopic size, he would squeeze so much mass into so small a volume that he would sink through the rocks of the crust to the center of the earth, a tiny speck of what we call "degenerate matter."

Throw out possibility one.

The second possibility would keep a man's complexity and density exactly right. But then what happens to the excess mass? The only thing that can possibly happen to disappearing mass (as far as we know) is to have it change into energy, and the shrinking man would thus become a super-powerful nuclear bomb.

What I did in the novel version of *Fantastic Voyage* was to make use of this second possibility and throw in a little vague analogy to the shrinkage of a photograph by the manipulation of three-dimensional optics. The reader could assume that a four-dimensional effect was involved with the excess mass disappearance. The mass went into hyperspace during the shrinking operation, I suppose, and came out of it again in the re-expansion.

This is fantasy, of course, but it shows, at least, that the problem exists. (In the movie, the matter of mass was entirely ignored.)

In the movie, and in the novel as well, the submarine is reduced to roughly the size of a large bacterium before being injected into the blood stream, so we can say that the human beings on board were 1/1000,000 centimeters tall (or 1/250,000 inch, if you prefer).

This means that if they were of average height to begin with, all their measurements have shrunk down to 1/17,-000,000 of what they were. To themselves, with their shrunken atoms and their diminished mass, they seem perfectly normal. They and their submarine seem, to themselves, to be of normal size while the entire universe outside the submarine has increased its measurements by 17,000,000 in every direction.

Consequently, the blood vessels are huge conduits, the white corpuscles are big enough to swallow a submarine whole without discomfort, and so on. The picture went that far, anyway, but what about atoms themselves?

An atom is about 1/100,000,000 centimeter in diameter. Increase its measurements 17,000,000 times in all directions and it becomes about ⅙ of a centimeter across. Since the important gases in the atmosphere (oxygen and nitrogen) are made up of molecules which each contain two atoms, the molecules in ordinary, unshrunken air would seem ellipsoids that are ⅓ of a centimeter across at their longest diameter—at least to the shrunken people.

This means that the unshrunken atoms and molecules would be *big enough to see* as far as those shrunken people were concerned. The movie, however, does not take that into account.

At one point, the submarine runs short of air and so it sticks a snorkel into the patient's lungs and fills up its tanks with fresh air. But the snorkel opening is not much larger in diameter than the air molecules in the patient's lungs. Can you imagine how long it would take to draw air through the snorkel under those conditions? A slow leak in a tire would be speedy in comparison.

What's more, once the ship is filled with unshrunken air, how do you get those huge molecules up your nostrils and into your own shrunken lungs? And what do you do with those molecules once they are in your lungs? Can your own shrunken red corpuscles handle them?

I didn't think of that till after I had finished the novel and I had to go back in a wild perspiration to revise several pages. I had to use a device to shrink some of the air in the patient's lungs *before* pulling it through the snorkel and into the ship.

Here's another point. The men on the shrunken submarine communicated with the outside world by radio. However, the radio was shrunken and the radio waves it emitted would have only 1/17,000,000 the wave length they would have had in the unshrunken state. The radio would be emitting light waves. They might seem like radio waves to the operator on board the submarine, but they would be tiny flashes of light to the men in the unshrunken world and using them for communication radio-fashion would be tricky.

And how would our men on the submarine see? By the light waves produced by their shrunken light sources? But these light waves would be gamma rays to the outside world—to the patient in whose blood stream they were cruising, for instance. Not enough to damage him, I hope, but I didn't bother to do any calculations.

I let the radiation bit go because (once again) I thought of it too late and was lazy enough to feel that no one would catch it. I underestimated my readers, of course. One sharp-minded young man picked it up and was down on me at once. I had to write back a confession of guilt.*

The movie-makers had the heroine (Raquel Welch) attacked by antibodies, but hadn't the faintest notion of what an antibody would look like if it were properly enlarged. Of course, with Miss Welch on screen, who studies the antibodies, anyway?

The antibodies are, of course, protein molecules and I imagined they would look like glimmering little balls of wool perhaps two inches across on the scale of the shrunken people. I had them feel like balls of wool, too, for the hydrogen bonds that held the peptide chains in place should be quite flexible and resilient.

Then, too, the movie-makers forgot to consider that the thin cell membranes would not be thin at all to the shrunken people. At one point, one of the men must work his way out of the capillary and into the lung. In the movie, there's no particular problem in doing so. You just slice through the paper-thin membranes separating the two. After all, membranes are only 1/10,000 of an inch thick.

* Since this article appeared, I received several letters pointing out additional subtle ways, involving electric forces, in which radical shrinking introduces insoluble dilemmas.

Sure, but to the shrunken people, the membrane would be something like 40 yards thick on their scale. The plot made it necessary for the hero to cross that thickness and I thought 40 yards was a bit much. I cheated and called it "several yards" in the book and let it go at that.

What's more, there was the matter of surface tension.

In the body of a quantity of liquid, each molecule is weakly attracted by all the other molecules. The attractions come from all directions and cancel each other out so that individual molecules move freely as though they are not being attracted at all.

At the surface of the liquid, a molecule is attracted only by other molecules within the liquid. The sparse scattering of air molecules outside the liquid has hardly any effect.

Molecules on the surface endure a net inward pull, therefore, and it takes an expenditure of energy for them to stay on the surface. For that reason, there is a tendency for the surface to be as small as possible. That is why small quantities of liquid, floating freely, assume a spherical shape. A sphere has the smallest possible area of surface for its volume. (A falling raindrop is "tear-shaped" because of air resistance.)

What's more, since all the surface molecules push inward as far as they can, they force themselves (so to speak) closer together, like people trying to push into a crowded subway car at the rush hour. To try to separate these crowded-together molecules takes more energy than trying to separate ordinary molecules in the body of the liquid. This extra clinging-together of surface molecules is spoken of as "surface tension" and it is as though the liquid had a very fragile skin covering its surface.

Tiny objects are not heavy enough to break that skin and small insects can go skittering over a water surface, not because they are floating (if they were within the body of water, they would *not* rise to the top) but because they are supported by the surface-tension skin.

Well, now, if insects are light enough to be supported by surface tension, what about objects the size of our shrunken people? I didn't dare calculate what the surface tension of unshrunken liquids might seem to be to the shrunken people. I suspect it would be so great that my heroes simply couldn't break through the surface of the liquid blood into the air within the lung.

The necessities of the movie plot forced me to let him through, albeit I made him have difficulties. So far no one has written in to say that the surface tension would be an insuperable barrier.

To explain one last point, let's go back to the nineteenth century, when there were indeed great scientists who, like Hollywood movie-makers, disbelieved in the existence of atoms. In the case of the scientists, it was not a matter of blissful ignorance, however, but of cogent thought.

The atomic theory had first been advanced in its modern form in 1803 by the English chemist John Dalton as a convenient way of explaining various chemical phenomena. Throughout the nineteenth century, the concept of atoms had been more and more successful in explaining what went on inside the test tube. By the end of the century, chemists were even making use of "structural formulas" for the more complicated molecules. Not only did they count the number of atoms of the various different kinds within a particular molecule; they even placed those atoms in specific arrangements in three dimensions, like a Tinkertoy structure.

Naturally, it was almost inevitable for chemists to believe that atoms really existed. If atoms did not exist how could their pretended existence explain so much so conveniently? How could matter behave *as though* it were atomic in so many ways and so thoroughly, if it were, in fact, *not* atomic?

Nevertheless, some scientists maintained it was not wise to stir beyond the measurable phenomena. All the nineteenth-century knowledge of atoms was indirect. Atoms were too small to be seen or sensed in any direct way, and while they might be very useful as a model to focus thought, they might (it was felt) mislead scientists who too easily believed in their literal existence.

The last great scientist to argue in this way, the last to

refuse to accept the literal existence of atoms, was the German physical chemist Wilhelm Ostwald. The twentieth century opened with Ostwald still vigorously maintaining the anti-atomic view. Yet he changed his mind, at last, and here's how:

It began in 1827 with a Scottish botanist, Robert Brown. Brown was interested in pollen and at one time he was studying a suspension of pollen grains in water under a low-power microscope. He found he had trouble focusing, for the grains jiggled. They did not move purposefully in a particular direction; they jiggled randomly all over the lot.

Brown realized that the pollen grains were alive, if dormant, and felt that the motion might be a kind of manifestation of life. However, he watched dye particles (indubitably nonliving) of similar size when suspended in water and they jiggled about erratically, too. Anything that was small enough would jiggle in suspension, and the smaller it was, the more marked was the jiggling.

All that could be done at this time was to give the phenomenon a name, since there was no explanation. It was discovered by Brown, and it also involved the erratic motion of particles. Why not call it "Brownian motion" then?

In the 1860s, the Scottish physicist James Clerk Maxwell produced an impressive explanation of the properties of gases in terms of randomly moving particles, and for the first time the atomic theory explained physical phenomena as well as chemical phenomena. At once the thought arose that randomly moving particles in a liquid might be shoving the larger particles in suspension this way and that.

In short, the grains of pollen or dye were being bombarded by water molecules and it might be that which was producing the Brownian motion.

Look at it this way. We ourselves are being bombarded from all sides by air molecules or, if we are in water, by water molecules. However, at any given moment, we are being struck by enormous numbers from all the different directions and the effect cancels itself out. We are struck by no more from one direction than from the opposite direction. It may be that a *few* more strike from the east than from the west, but the individual molecules are so tiny that the effect of a mere few out of the astronomical numbers available is immeasurably small. Of course, if a *lot* more strike from the east than from the west, we will feel the push, but the chance that a *lot* more will do so in the sheerly random motion of the molecules is also immeasurably small.

But consider our shrinking people (or shrinking anything). From their point of view the universe is getting larger, and the molecules of air and water as well. Because they themselves get smaller, they present a smaller target to the bombarding molecules and fewer strike them in a given small instant of time. The individual molecules get larger and larger, too.

When our people reach the microscopic, they are being bombarded by BB-shot and not by many of them. *Now* if a few more strike from east than from west it will be important and their tiny bodies will feel it. A preponderance from the east shoves them west, and in the next moment a preponderance from above will push them downward, and so on.

The random bombardment by molecules would explain the motion in the first place, the erratic nature of the motion in the second, and the fact that the effect was more pronounced the smaller the floating object in the third.

Men like Ostwald were not impressed by this argument. It was just talk. To make it more than talk one ought to calculate the chances of imbalance in molecular bombardment and the size of the effects. In short, there would have to be a strict and rigorous mathematical analysis of Brownian motion that would explain it *quantitatively* in terms of the random bombardment of molecules.

Then in 1905, such an analysis was produced, and by none other than Albert Einstein.

According to his equation, particles suspended in a tall container of liquid ought to behave in response to a balance between the force of gravity and the effect of Brownian motion.

Gravity pulled downward and molecular motion kicked in all directions, including upward. If gravity were all that were involved, the particles would all settle to the bottom. If Brownian motion were all that were involved they would spread out evenly. In a combination of the two, they would spread out, but pile up more and more densely toward the bottom. The more massive the molecules, the greater the Brownian motion for objects of a particular size at a particular temperature and the smaller the extent to which they would crowd toward the bottom.

Einstein was merely a theoretician, however. He was satisfied to produce the equation and he left it to others to check it against observable phenomena. The one who followed it up was a French physicist, Jean Baptiste Perrin. In 1908, he suspended small particles of gum resin in water and counted the number of particles at different levels. He found that the numbers increased as he worked downward exactly in accordance with Einstein's equation, provided the molecules of water were given a certain mass.

Not only had he verified Einstein's explanation of Brownian motion, but he was the first to work out a reasonably accurate measure of the actual weight of individual molecules.

Ostwald was now faced with an observable effect produced by individual molecules. By looking at suspended objects in water jiggling around, he could, in effect, see them being struck by individual molecules. That supplied his rigid requirement for direct observable evidence and he could no longer deny the existence of atoms.

Perrin received the 1926 Nobel Prize in Physics for his work. (Einstein had gotten his, for other work, in 1921.)

This brings me back to *Fantastic Voyage*. In the movie, the shrunken submarine moves through the blood stream precisely as it would if it were of normal size moving through an ocean current. That would be so if there were no atoms or if atoms were infinitesimally small.

But there *are* atoms of size comparable to the shrunken ship. The ship ought therefore to be jiggling from side to side, back and forth, up and down, and every direction in between, through the effect of Brownian motion. This would happen in even more pronounced fashion to the individual people on those occasions when they left the ship to swim around in the blood stream.

I couldn't actually allow much Brownian motion in the novel. It would have introduced too many complications (bad seasickness for one thing). I did mention it, though, and allowed some jiggling at the start, just to show I knew it ought to be there; and then I ignored it.

All this may cause alarm and despondency in the hearts of some of my Gentle Readers who have felt the impulse to write science fiction and who now think it requires advanced degrees in science to do so. Don't think that for a moment. Good science fiction does not necessarily require all this folderol.

Despite all the errors, inconsistencies, and oversights in the movie *Fantastic Voyage*, I thought it was a lot of fun, very suspenseful, fast-moving, and delightful. The errors didn't bother me a bit while I was watching.

What's more, if someone else had written a novel based on the movie and had not bothered to correct any of the errors, it would still easily have made an exciting story.

It's just that, in my own case, I happened to know the errors were there and I had to correct them. It's my personal way of writing science fiction and it's not the only way. Candor compels me to say that I consider my way the best way, but it's still not the only way.

B CHEMISTRY

II The First Metal

I am sometimes asked how I decide on the subject for a science essay. The answer is clear and straightforward: I don't know.

Sometimes, though, I do happen to catch a fugitive glimpse of the mental processes involved before they whiff away and depart forever.

Thus, several weeks ago, I came across some comments in a chemistry journal concerning the metal, gallium. This is a very interesting metal on two counts. It played a melodramatic role in the working out of the periodic table, and it has a very interesting melting point.

That gave rise to the possibility of an essay on the periodic table or, alternatively, to one on melting points of metals. For a few moments, I speculated idly on what could be done with melting points. It seemed to me that if I were going to discuss the melting point of gallium, I would first have to discuss the melting point of mercury.

And if I discussed the melting point of mercury, I would have to mention a few other things about it, notably the fact that it was one of the seven metals known to the ancients.

159

In that case, how about an essay, first, on the ancient metals? That is what I am now sitting down to write, intending to work my way up to mercury, and then to gallium.

And that is how I decide on subjects to write about—at least, in this case.

The seven metals known to the ancients were (in alphabetical order): copper, gold, iron, lead, mercury, silver, and tin. The discovery of each of these is lost in the mists of the past, but I strongly suspect it was gold that was discovered first. It was gold that was the first metal.

Why not? Gold can occasionally be found as a glittering little nugget. Its bright and beautiful yellow color would easily attract the eye and it would become an ornament at once.

Once handled, gold would almost at once show itself to be a remarkable substance, much different from the rock, wood, and bone that mankind had been working with for hundreds of thousands of years. Not only would it have a shiny color, but it would be considerably heavier than an ordinary pebble of the same size.

Then, too, suppose the finder wished to work the nugget into a more symmetrical shape. To shape a stone, he would make use of careful strokes with a stone chisel. Flakes of thin stone would split off the object being shaped.

The gold would not behave in this manner. The chisel would merely make a dent in the gold. If the gold were beaten with a mallet, the metal would not powder as a pebble would; it would flatten into a very thin sheet. It could also be drawn out into a very fine wire, which stone certainly could not do.

Other metals were eventually discovered—other objects which had luster and unusual weight and which were malleable and ductile. None was as good as gold, though. None was as beautiful, or as heavy. What's more, other metals tended to lose their shine more or less quickly if exposed to air over periods of time; gold never did.

And gold had another property which added to its

value; it was rare. This was true, to a somewhat lesser extent, however, of the other metals, too. The Earth's crust is primarily rock, and the occasional metal nugget was occasional indeed. The very word "metal" seems to come from the Greek word *metallan*, meaning "to search for," a tribute both to its rarity and desirability.

Modern chemists have worked out the composition of the Earth's crust in terms of each of the various elements, including the seven ancient metals. Here in Table 32 are the figures for the seven, given in grams of metal per tons of Earth's crust, and in order of decreasing concentration.

TABLE 32

Metal	Concentration (g/ton)
Iron	50,000
Copper	80
Lead	15
Tin	3
Mercury	0.5
Silver	0.1
Gold	0.005

As you see, gold is by far the rarest of the seven metals. A concentration of 0.005 gram per ton is equivalent to 1 part in 200,000,000.

Still, the total quantity of gold is sizable if one considers the entire crust. At this percentage, the total mass of gold there is about 155,000,000,000 tons.

There is gold in the ocean, too, in the form of submicroscopic metallic fragments, which comes to a concentration of 0.000005 gram per ton, making a total gold mass in the oceans of nearly 9,000,000 tons.

The gold in the ocean is so dilute that it cannot yet be extracted at anything but a large loss. Therefore no gold has ever been extracted from the ocean. The soil has a larger concentration of gold, but the soil is harder to work with. If gold were evenly spread throughout the crust, that gold, too, would be unavailable to us.

But the gold is not evenly spread. There are occasional

accessible regions where the gold content is high enough to be mined profitably, even with primitive equipment, and where reasonably pure metallic gold can sometimes be found in sizable nuggets.

But only a tiny fraction of all the gold is thus made available. Nothing has been so avidly searched for as gold through all six thousand years of civilized history; yet with all that search it is estimated that the total amount of gold extracted from the ground by mankind amounts to only 50,000 tons. What's more, the world's mines are producing gold at the rate of only about a thousand tons a year (half of it in South Africa). Even so, the end of Earth's reserves of minable gold seems to be in sight.

It would be interesting to see how small a quantity of gold has served to affect human history in so enormous a way. If all the gold so far extracted from the Earth were packed into a cube, that cube would be 290 feet on each side. If all that gold were used to plate an area the size of Manhattan Island, the layer of gold would only be about 1 millimeter thick. (That puts another view on the old immigrant notion that the streets of New York are paved with gold. It would have to be, at best, an awfully thin pavement.)

The question then arises as to why gold should have been the first metal to be discovered if it was the rarest of the seven.

The answer lies in the comparative activity of the metals, their comparative tendency to combine with other elements to form nonmetallic compounds.

The activity of metals can be measured as "oxidation potential" in volts (because electric currents can make metallic atoms plate out as free metals or go into solution as ions). The element hydrogen (which has some metallic properties from a chemical standpoint) is arbitrarily given an oxidation potential of 0.0 volts. Elements which are more active than hydrogen have a positive oxidation potential; those less active, a negative one.

Here, then, in Table 33, are the oxidation potentials for the seven ancient metals:

TABLE 33

Metal	Oxidation Potential (volts)
Iron	+0.44
Tin	+0.14
Lead	+0.13
Copper	−0.34
Mercury	−0.79
Silver	−0.80
Gold	−1.50

As you see, gold is by far the least active of the seven metals, and is therefore by far the most likely to exist in free metallic form. Thus, gold is far less common than iron, atom by atom, but gold nuggets are far more common than iron nuggets. Indeed, but for one factor which I will come to soon, iron nuggets would not be found at all. Furthermore, the yellow gleam of gold is much more likely to be noticed than the dirty gray of iron.

So it happens that while silver and copper objects (also among the more inactive metals) can be found in predynastic Egyptian tombs dating as far back as 4000 B.C, gold objects, it is thought, antedate that mark by several centuries.

In early Egyptian history, silver was more expensive than gold, simply because it was rarer in nugget form.

Indeed, we might generalize that the ancient metals are the inert metals. Yet we are bound to ask, then, if there were any inert metals *not* known to the ancients. The answer is: Yes.

There are six metals of the "platinum group"—platinum itself, then palladium, rhodium, ruthenium, osmium, and iridium—that should qualify. Platinum, osmium, and iridium are somewhat more inert than gold, even, and the others are at least as inactive as silver. Why, then, were they not known to the ancients?

It is tempting to blame it on the rarity of the metals. Four of them—ruthenium, rhodium, osmium, and iridium —are considerably rarer even than gold, with concentrations in the crust of only 0.001 gram per ton. They, along with rhenium, have the distinction of being the least common metals on Earth. (Rhenium has also the unique distinction of being the last stable element to be discovered—but that is another story.)

Yet platinum is as common as gold, and palladium twice as common. If gold nuggets can be found, then, why not nuggets of platinum? Or palladium?

For one thing, yellow gold is far more noticeable than white platinum. For another, the best platinum ores are located nowhere near the ancient sites of civilization in the Middle East.

Then, too, I rather suspect that platinum nuggets were indeed found now and then—and mistaken for silver. Platinum is far less malleable than silver and is not easily worked. I can see the primitive metalworker looking at such nuggets in disgust and muttering "Spoiled silver" as he tosses them away.

The apparent resemblance to silver marks platinum to this day. It was first clearly recognized as a distinct metal in 1748, when a Spanish chemist, Don Antonio de Ulloa, described samples of the metal which he had located in the course of his travels through South America. He named it "platina" from *plata*, the Spanish word for silver. So platinum remains forever (at least in name) a kind of silver.

Nor is it surprising, in view of all this, that iron, by far the most common of the seven metals—five hundred times as common as all the remaining six put together—lagged in some ways behind the rest. After all, it was the most active of the ancient metals, the most apt to be in combination, the most difficult to loose from that combination.

That it was known at all may have been the result of a cosmic catastrophe millions of miles from Earth.

After all, as far as chemical principles are concerned, iron should occur on Earth only in the form of nonmetallic compounds, never as the free metal. Yet that's not the way it is.

There is so much iron on Earth, and it is so concentrated toward the center, that one-third of the mass of the planet is a liquid core of iron plus its sister metal, nickel, in a 10-to-1 ratio. In itself, this doesn't affect Earth's crust, but there must be other planets with such a nickel-iron core and it seems that one of them may have exploded (the one, presumably, between Mars and Jupiter, the orbit of which is now marked by the asteroids, the fragments of that explosion). The smaller fragments of that explosion bombard the earth, and some of them are the fragments of the nickel-iron core. If the fragment is large enough it survives the friction of the atmosphere and strikes the crust, where it lodges as heaven-born "nuggets" of iron.

Small pieces of nickel-iron (undoubtedly meteoric in origin) are found in Egyptian tombs dating back to 3500 B.C. They are there in the guise of jewelry.

As long as metals could only be used when found as nuggets, they were bound to be excessively rare, but some time before 3500 B.C. the true discovery of metals was made. Any fool, after all, could stumble across a nugget. It took a man, however, to recognize what had happened when copper nuggets were found in the ashes of a fire that had been built on a blue stone.

It was a daring thought that from rock one might gain metal. The science of metallurgy began and men began to look not for metal only, but for metal ores—for rocks that, on heating in a wood fire, would yield metal.

It was copper that was chiefly produced in this manner, and it became the wonder metal of the age. It was 16,000 times as common as gold, once the ores were taken into account. And though it did occur as rocklike compounds in the ore, it was not so active as to be tightly bound in those compounds. A gentle nudge, chemically speaking, would suffice to pry the copper loose.

Copper alone was suitable merely for ornaments and for some utensils—too soft for anything else. But then another accidental discovery must have been made. Tin ores could be handled much as copper ores were, and if some ores contained both copper and tin, the mixed metal ("alloy") that resulted was much harder and tougher than copper alone. We call the alloy "bronze." The ancients learned to mix ores of copper and tin on purpose and to use the bronze they obtained for war weapons. Thus was initiated the Bronze Age. In the Middle East, the site of man's oldest civilizations, the Bronze Age began about 3500 B.C. and endured for something over two thousand years.

Tin was the bottleneck here. It is only 1/25 as common as copper, and the tin reserves of the Middle East ran out while copper was still available in comfortable amounts. As a result, the far-distant corners of the world had to be scoured for tin. The Phoenician navigators, the best and most daring of the ancient world, made their way out to the "Tin Isles" for the purpose.

The Phoenicians kept the secret of the location of the Tin Isles throughout their history, but it seems quite certain that they were sailing out into the Atlantic Ocean and northward to Cornwall, on the southwestern extremity of the island of Great Britain.

Cornwall is one of the few regions of the Earth which is rich in tin ore. In twenty-five centuries of steady mining, some 3,000,000 tons of tin have been removed from the Cornish mines and the area is not yet exhausted. Nevertheless, its output these days is minute compared to those of the relatively untapped mines of Malaysia, Indonesia, and Bolivia.

Yet even while bronze was carrying all before it, the ancients knew very well that there was a metal that was harder and tougher than bronze and potentially much better for war weapons and tools. It was iron—those metallic lumps that were picked up now and then, a very rare now and then.

There were, of course, iron ores, just as there were copper ores and tin ores. Indeed, it was obvious that iron ores were extremely common. The trouble was that iron (much more active than copper) held on to its place in the compound firmly. The techniques that sufficed to extract metallic copper would not do for iron. Such iron as was coaxed out of the ore was riddled with gas bubbles. It was brittle and good for nothing.

Special techniques involving particularly hot flames were needed, along with high-quality charcoal. Even when temperatures were reached that sufficed to melt the iron and drive out the bubbles and, indeed, prepare it in pure form, the end-product was disappointing. Iron obtained from ore was not nearly as hard as the meteoric nuggets, and it did not hold nearly as fine an edge. The difference was that meteoric iron contained nickel (a metal unknown to the ancients).

But then processes were developed that produced iron into which some carbon from the charcoal was introduced. In effect, a kind of steel was produced and that, at last, was the metal that was needed.

It was sometime about 1500 B.C. when the secret of producing good iron in useful quantities was developed somewhere in the southern foothills of the Caucasian mountains. This was in what was known as the kingdom of Urartu (the "Ararat" where Noah's Ark landed). The area was, at the time, under the control of the Hittites, whose power center was in eastern Asia Minor. The Hittite kingdom tried to keep knowledge of the new technique a monopoly, but the exploitation of this new weapon was slow. Before the Hittites could really turn the metal into a world-beating military resource, they were themselves beaten by a combination of civil war and invasion from without.

The fall of the Hittites came soon after 1200 B.C. and the secret of iron technology fell to Assyria, the land just south of Urartu. Gradually, the Assyrians developed iron to an unprecedented extent and by 800 B.C. they were putting a completely iron-ized army into the field. They stockpiled iron ingots as we stockpile uranium and for a similar purpose. For two hundred years, the Assyrians swept all before them and built the greatest empire the Middle East had yet seen—until their victims learned iron technology for themselves.

It is interesting to note, by the way, that iron, despite its commonness, is not the most common metal on Earth. There is one metal more common, but also more active.

In consequence, it lagged even farther behind in development.

The most common metal in Earth's crust is aluminum, the concentration of which is 81,300 grams per ton. It is 1.6 times as common as iron, but its oxidation potential is +1.66, which is considerably higher than even that of iron.

This means that the tendency of aluminum to form compounds is still greater than that of iron, and that it is much more difficult to force aluminum out of those compounds than it is to force iron out of its compounds. Moreover, no aluminum nuggets fell from the sky to hint to mankind that such an element exists.

As a result, aluminum remained completely unknown (as a free metal) to the ancients. It wasn't until 1825 that the first piece of aluminum metal (quite impure) was forced out of a compound by the Danish chemist Hans Christian Oersted. And it wasn't till 1886 that a good method was discovered for producing the pure metal cheaply and in quantity.

Metals, generally, are denser than stone. If we measure density in ounces per cubic inch, we find (see Table 34):

TABLE 34

Metal	Density (ounces/cubic inch)
Tin	4.2
Iron	4.6
Copper	5.2
Silver	6.1
Lead	6.6
Mercury	7.9
Gold	11.3

Since the typical rock has a density of about 1.6 ounces per cubic inch, even the least dense of the seven metals is

2.5 times as dense as rock, while gold is about seven times as dense.

High density has its uses. If you wish to pack a lot of weight into a small volume, you would use metal rather than stone; the denser the metal the better. Gold is the best in this respect but no one is going to use gold as routine ballast; it is too valuable. Mercury, being liquid, would be too hard to handle.

That leaves lead as a third choice. It is relatively cheap, for a metal, and it is four times as dense as rock. Lead, therefore, became the representative of heaviness. The phrase "heavy as lead" has entered the language as a cliché, which has much more force, through sheer repetition, than the phrases "heavy as gold" or "heavy as platinum." (We mean "dense" rather than "heavy," but never mind.)

Again, we speak of "leaden eyelids" to imply sleepiness that can't be fought off; "leaden steps" to indicate a walk made slow and difficult by the weight of weariness or sorrow.

To force a line to hang vertically, one would want to put a weight at one end, so that the force of gravity would stretch the line straight up and down. A lump of lead would be a compact way of supplying the weight.

The Latin word for "lead" is *plumbum* and now you can see what a "plumb line" must be. Since you would attach a piece of lead to a line you wanted to throw into the ocean and have sink as far as possible, you see what "to plumb the depths" means.

Again, since the ancients believed that the heavier an object the faster it fell, it seemed to them that a lead weight would fall faster than the same-sized weight made of other less dense materials. So you see what "to plummet downward" means.

Finally, since a lead weight makes a line *completely* vertical, there grew to be a tie-in between lead and completeness, so that now you see why, in a Western movie, the old rancher says to the young schoolmarm, "By dogies, Ah'm shore plumb tuckered out, missie." (At least you know why he says it if you know what the other words mean.)

All this remains, even though, in addition to gold and mercury, six other metals, now known, are denser than lead. Three of these newcomers—platinum, osmium, and iridium—are denser, even, than gold. Osmium has a density of 13.1 ounces per cubic inch and the other two are not far behind.

One more point. Metals, generally, are lower-melting than most rocks. Rocks melt, in general, at temperatures of 1800° to 2000° C. This is high enough to allow rocky materials to be used to construct furnaces and chimneys. Here in Table 35 are the melting points of the seven ancient metals:

TABLE 35

Metal	Melting Point (° C.)
Iron	1535
Copper	1083
Gold	1063
Silver	961
Lead	327
Tin	232
Mercury	—39

Iron has quite a high melting point for a metal, which is one of the reasons it was a hard nut, metallurgically, for the ancients to crack. Copper, silver, and gold are in the intermediate range, but look at lead and tin.

Lead and tin are easy to melt in any ordinary flame, and a mixture of the two will melt at lower temperatures than either separately—temperatures as low as 183° C. Such an alloy of tin and lead is "solder." It can be easily melted, poured onto the joint between two pieces of metal, and allowed to freeze.

Tin containing a little lead is "pewter." Kings and nabobs used gold and silver plates to eat from, despite their expense and difficulty of working, out of a sense of conspicuous consumption. Poor people used clay and wood, which were ugly. In between, there were pewter dishes.

It was particularly easy to work either tin or lead into tubes, and there is a tale to tell about each. Ordinary metallic "white tin" is stable only at relatively warm temperatures. In winter cold there begins to be a tendency for it to turn into a crumbly nonmetallic "gray tin." This takes place slowly unless temperatures considerably below zero are reached.

A cathedral at St. Petersburg, Russia, installed a magnificent organ with beautiful tin pipes. Came a cold, cold winter and the pipes disintegrated. That's how chemists discovered about white tin and gray tin, but I doubt that the cathedral personnel were particularly overjoyed at their contribution to scientific knowledge.

It is all very well to build organ pipes out of tin, but for ordinary plebeian water pipes, tin was too expensive. The other low-melting metal, lead, was used. In those parts of the Roman Empire where a central water supply was set up (in the city of Rome itself, for instance), lead pipes were used. And now you know why we call the fellow who deals with water pipes a "plumber," even though the pipes are no longer made of lead.

As it happens, and as the Romans did not know, lead compounds are strongly and cumulatively poisonous. Under certain conditions, small quantities of the lead pipe would dissolve and the water supply would become dangerous over longish periods.

There are even some who have recently suggested that the Roman Empire fell, in part at least, because key men of the government and social leadership in the city of Rome were suffering from "plumbism," that is, from chronic lead poisoning.

But neither silver nor lead is the lowest-melting of the

seven ancient metals. That record was held (and is still held today) by mercury—and that brings us back one step in the chain of rumination I described at the start of this chapter.

Mercury will be the subject of the next chapter.

12 The Seventh Metal

It is very difficult for an ivory-tower chemist such as myself to demonstrate competence in the practical aspects of the science. Consequently, it is always with a sinking heart that I watch anyone approaching me with a down-to-earth problem in chemistry. It always ends in a personal humiliation.

Well, not always.

Once, in the days when I was working toward my Ph.D., my wife came to me in alarm. "Something," she said, "has happened to my wedding ring."

I winced. I was still in my early stages as a chemist, but I had already had time to demonstrate my incompetence many times over. I didn't enjoy the prospect of having to do so again.

I said, "What happened?"

She eyed me censoriously, "It's turned into silver."

I stared at her with astonishment. "But that's impossible!"

She handed me the ring and, indeed, it had the appearance of silver, yet it was her gold wedding ring, inscription and all. She waited and I felt, uneasily, that she suspected I had bought her a ring of low-grade gold. Yet I could think of nothing!

I said, "I just can't explain this. Except for mercury, there's nothing in the world—"

"Mercury?" she said, with rising inflection. "How did you know about the mercury?"

I had apparently said the magic word. I saw instantly what had happened. Inflating my chest and putting on an air of lofty consideration, I said, "To the chemist's eye, my dear, it is at once obvious that what one has here is gold amalgam and that you've been handling mercury without removing your wedding ring first."

That was it, of course. At the laboratory I had had access to mercury and was fascinated by it so I brought home a small vial of it to amuse myself with now and then. (It rolls around in enticing fashion and doesn't wet anything.) My wife found the vial and couldn't resist pouring a drop into the palm of her hand so *she* could play with it. But she kept her wedding ring on and mercury rapidly mixes with gold to form a silvery gold amalgam.

Yet despite this sadly dramatic and highly personal instance of mercury's fascination, I discussed the seven metals known to ancient man in the preceding chapter with hardly a word for the most unusual of them— mercury. But that wasn't neglect; I was merely saving it for a chapter of its own.

Mercury is riddled with exceptional characteristics. I am sure, for instance, that it was the least familiar of the seven metals and strongly suspect that it was the seventh metal (and the last) to be discovered by the ancients.

As for its being the least familiar, we might see what the Bible has to say about it, if only because it is a long and intricate book written by people who had little or no interest in science. It might therefore be considered the authentic voice of the ancient nonscientist.

Gold, of course, is the standard of excellence and perfection to all, even to the Biblical writers. To say something is more than gold is to give it the highest possible worldly praise. Thus:

> Psalms 119:127. *Therefore I love thy commandments above gold; yea, above fine gold.*

And as nonscientists, what do the Biblical writers say about the other metals? For economy's sake, I've searched for a verse that mentions as many metals as possible, and here's one where Ezekiel is quoting God as threatening the sinners among the Jews:

> Ezekiel 22:20. *As they gather silver, and brass, and iron, and lead, and tin, into the midst of the furnace to blow the fire upon it, to melt it; so will I gather you in mine anger and in my fury, and I will leave you there, and melt you.*

The sinners are compared to the various metals, notably excluding gold, to show that they are imperfect.

Here, by the way, we must remember that the English words of the King James version are translations of the original Hebrew and may be mistranslations. The Hebrew word *nehosheth* was used indiscriminately for pure copper and for bronze, an alloy of copper and tin. The King James version invariably translates it as "brass," which is an alloy of copper and zinc and is *not* what the Biblical writers meant. The Revised Standard Version replaces all the "brass" in the King James with "bronze" or "copper."

If we substitute copper for brass in the verse from Ezekiel, you will see that I have managed, by using merely two Biblical verses, to mention six of the ancient metals: gold, silver, copper, iron, tin, and lead. That leaves only mercury. What does the Bible say about mercury?

The answer is: Nothing!

Not a word! Not in the Old Testament, or the New, or the Apocrypha. It seems clear that of all the seven metals, mercury was the most exotic, the least used for everyday purposes, the most nearly what we would today call a "laboratory curiosity." The nonscientists who wrote the Bible were so little acquainted with it, they never had reason to mention it, even in figures of speech.

As to why it should be the last to be discovered; that seems to me to be no mystery. It is comparatively rare, for of the seven metals, only silver and gold are less common. Nor does one come across mercury ingots, naturally, since it is liquid at ordinary temperatures.

What led to its discovery was the accident that its one important ore was brightly colored. This ore is "cinnabar," which, chemically speaking, is mercuric sulfide, a compound of mercury and sulfur. It has a bright red color and can be used as a pigment. When so used, it is called vermilion, a word also applied to its color.

Cinnabar must have been in considerable demand and, undoubtedly, there must have been occasions when it was accidentally heated to the point where it broke up and liberated drops of metallic mercury. There is evidence in the Egyptian tombs that mercury was known in that land at least as far back as 1500 B.C. This sounds ancient enough but compare it with copper, silver, and gold, which date back to 4000 B.C.

Even after mercury had been isolated, there seems to have been difficulty in recognizing it as a new and separate metal. The fact that it was liquid may have made it too different from the other metals to put on a par with them. Perhaps it was only one of the other metals in molten form.

It had the look of silver about it. Could it be, then, liquid silver? Silver itself, ordinary solid silver, could be melted if raised to a good red heat, but mercury was a liquid silver at ordinary temperatures. To the ancient workers, such a difference was perhaps not as significant as it would be to us. If there could be a hot liquid silver, why not a cold one?

In any case, whatever the thought processes of the early discoverers of mercury, it remains true that mercury was the only one of the seven metals not given a name of its own. Aristotle called it "liquid silver" (in Greek), and in Roman times the Greek physician Dioscorides called it "water-silver," which is essentially the same thing. The latter name is *hydrargyros* in Greek and became *hydrargyrum* in Latin. And in fact, the chemical symbol for mercury remains Hg, in honor of that Latin name, to this day.

The Roman writer Pliny called it *argentum vivum*, meaning "living silver." The reason for this is that ordinary silver was solid and motionless (that is, "dead") whereas mercury quivered and moved under slight impulse. If a drop fell, it shivered, and the droplets darted away in all directions. It was "alive."

An old English word for alive was "quick." We still use it with that meaning in the old phrase "the quick and the dead." We still say that vegetation "quickens" in the spring. If we cut past the outermost dead callus of the skin, to the soft, sensitive tissue beneath, we "cut to the quick."

Naturally, "quick" comes to be applied to the more notable characteristics of life, one of which is rapid motion. To be sure, there are forms of life, such as oysters, sponges, and mosses, which don't show notable motions, but the language-making commonalty indulged in no such fine side-issues. They knew the distinction between a race horse and a hobbyhorse. Consequently, "quick" came to mean "rapid."

Nowadays, that last meaning of "quick" has drowned out everything else, and the older meaning remains only in the old clichés that never die and that remain to puzzle innocents. (Moderns would imagine that "the quick and the dead" refers to Los Angeles pedestrians.)

Keeping all this in mind, we see why Pliny's *argentum vivum* can be literally translated as "quicksilver," and that is, indeed, the old English name for it.

Where, then, did the name "mercury" come from?

The medieval alchemists approached their work in a thoroughly mystical manner. Since most of them (not all!) were incompetent, they could best mask their shortcomings by indulging in windy mysteries. What the public could not understand, they could not see through.

Naturally, then, alchemists favored metaphorical speech. There were seven different metals and there were also seven different planets, and surely this could not be coincidence, could it? Why not, then, impressively speak of the planets when you meant the metals?

Thus, the four brightest of the planets, in order of decreasing brightness, were the Sun, the Moon, Venus, and Jupiter. Why not match these with gold, silver, copper, and tin respectively, since these were the four most valuable metals in order of decreasing value?

As for the others—Mars, the ruddy planet of the war god, is naturally iron, the metal out of which war weapons are made. (As a matter of fact, the ruddiness of Mars may be due to the iron rust in its soil. It's this sort of coincidence that causes modern mystics to wonder if "there might not be something in alchemy." To counter that, one need only say that any random succession of syllables is bound to make words now and then and if you carefully select out the words and don't touch anything else, you can easily convince yourself that nonsense is sense.)

The slow-moving Saturn, slowest of all the planets, is naturally matched to lead, the proverbial standard for dullness and heaviness. Mercury, on the other hand, which swings rapidly from one side of the Sun to the other, is equated with the darting droplets of quicksilver. (Thus, the seventh metal is matched with the seventh planet. See Chapter 1. Pure coincidence!)

Some of these comparisons still hang on in the form of old-fashioned names for certain compounds. Silver nitrate, for instance, appears in old books as "lunar caustic" because of the supposed relationship of silver and the Moon. Again, colored iron compounds used as pigments are sometimes called by such names as "Mars yellow" or "Mars red." Lead poisoning was once referred to as "saturnine poisoning."

The only planet to enter the realm of modern chemistry in a respectable way was Mercury. It became the name of the metal, ousting the older quicksilver. Perhaps this came about because chemists recognized that quicksilver was not an independent name and that mercury was not merely silver that was liquid or quick.

Oddly enough, metals were named for planets in modern times, too, and the modern names stuck, of course. In 1781, the planet Uranus was discovered, and in 1789, when the German chemist Martin Heinrich Klaproth discovered a new metal, he named it for the new planet and it became "uranium." Then, in the 1940s, when two metals were found beyond uranium, they were named for two planets, Neptune and Pluto, that had been found beyond Uranus. The new metals became "neptunium" and "plutonium."

Even the asteroids got their chance. In 1801 and 1802, the first two asteroids, Ceres and Pallas, were discovered (see Chapter 4). Klaproth discovered another new metal in 1803 and promptly named it "cerium." The same year, an English chemist, William Hyde Wollaston, discovered a new metal and named it "palladium."

Mercury gained unusual distinctions during the Middle Ages. Throughout ancient and medieval times, the chief source of mercury was Spain, and the Moorish kings of the land made spectacular use of it. Abd ar-Rahman III, the greatest of them, built a palace near Cordova about 950, in the courtyard of which a fountain of mercury played continuously. Another king was supposed to have slept on a mattress that floated in a pool of mercury.

Mercury gained another medieval distinction of a more abstract nature. It seems that one of the chief aims of most medieval alchemists was the conversion of an inexpensive metal like lead to an expensive one like gold.

That this could be done seemed likely from the old Greek notion that all matter was made up of combinations of four basic substances, or "elements," which were called "earth," "water," "air," and "fire." These were not identical with the common substances we call by those names, but were abstractions which might better be translated as "solid," "liquid," "gas," and "energy." It was not really a bad guess for the times.

The medieval alchemists went beyond the Greek notions, however. It seemed to them that metals were so different from the ordinary "earthy" substances, like rocks, that there must be a particular metallic principle involved. This metallic principle, plus "earth," made a metal. If one could but locate the metallic principle, one could add "earth" in different ways to form any metal, including gold.

Naturally, by adding "earth" to the metallic principle, one added solidity to it and produced a solid metal. What about mercury, then? It was a liquid and that must be because it had so little "earth" in it. Perhaps what little it did have could be removed in some fashion, leaving the metallic principle itself.

Many alchemists began to work with mercury indefatigably, and since mercury vapors are cumulatively poisonous, I wonder how many of them died prematurely. The vapors affect the mind, too, but I suppose it's hard to tell when an alchemist is speaking *real* gibberish. For that matter, I wonder about that Moorish king who slept over a pool of mercury. How did *he* feel as the months wore on?

Some alchemists must have reasoned further that gold was unique among metals for its yellow color; therefore, what must be added to mercury (itself silvery in color) is a yellow "earth." The obvious candidate for a yellow "earth" is sulfur. Sulfur was unusual in that, unlike other earths, it could burn, producing a mysterious blue flame and an even more mysterious choking odor. It seemed easy to seize on the idea that mercury and sulfur represented the principles of metallicity and inflammability respectively. The combination of the two would therefore put fire and solidity into mercury and turn it from a silver liquid to a golden solid.

To be sure, mercury and sulfur did combine—to form cinnabar. This was a perfectly ordinary red "earth," nothing like gold, but the dullness of fact was rarely allowed to spoil the glorious alchemical vision.

These medieval theories slowly died in the course of the eighteenth century, when real chemistry passed through its lusty infancy. During that century, the role of mercury as a metallic principle received a cruel knock on the head. As such a principle it would have to be a perpetual liquid, but was it?

The year 1759 was a very cold one in St. Petersburg, Russia, and on Christmas day there was a blizzard and the mercury sank very low in the thermometers. The Russian chemist Mikhail Vassilievich Lomonosov tried to get the temperature to drop still lower by packing the thermometer in a mixture of nitric acid and snow. The mercury column dropped to -39° C. and would drop no lower. It had frozen! The world, for the first time, saw solid mercury, a metal like other metals.

By that time, though, mercury had gained a new value that far outweighed its false role as a metallic principle. In a way, this new value was based upon its density, which is 13.6 times that of water. A pint of water weighs roughly a pound; a pint of mercury would weigh about 13½ pounds.

This is an amazingly high density. Not only would steel float in mercury, *lead* would do so. Somehow we don't expect this of a liquid; too much of our experience is with water. Thus, when a young chemistry student is brought face to face with his first sizable jug of mercury, he can be spectacularly astonished. If he is asked, casually, to pick it up and put it somewhere, he puts his hand around it and automatically gives it the kind of lift he would give a jug of water of corresponding size. And of course the mercury acts as though it were nailed to the table.

In 1643, the Italian physicist Evangelista Torricelli made use of mercury's density. He was puzzling over the problem that a pump could only lift a column of water 34 feet above its natural level. He reasoned that the actual work of raising that column was done by the pressure of the atmosphere. A column of water 34 feet high exerted a pressure at its base equal to the full pressure of the air so the water could be raised no higher.

To check that more conveniently (a 34-foot column is a clumsy thing to handle), Torricelli made use of mercury, the densest liquid known. A column of mercury (13.6

times as dense as water) would produce as much pressure at its base as a column of water 13.6 times as high. If 34 feet of water balanced the total air pressure, then 2.5 feet (or 30 inches) would also balance it.

Torricelli therefore filled a yard-long tube with mercury and upended it in a bowl of mercury. The mercury began pouring out, but only so far. When the height of the column had decreased to 30 inches, it then stayed put, balanced by the air. Torricelli had demonstrated his point and had invented the barometer. Mercury entered a new career as a unique substance (a very dense, electricity-conducting liquid) adapted to use in numerous instruments of science.

Incidentally, if the air were as dense all the way upward as it is at Earth's surface, we could easily calculate what the height of the atmosphere would be. Mercury is 10,560 times as dense as the surface air. Therefore, a column of mercury would balance 10,560 times its own height of air. This means that 30 inches of mercury would balance 5 miles of air.

The air, however, is *not* evenly dense all the way upward. It grows less dense as we rise and therefore bellies upward to great heights.

Of all the metals known to the ancients, mercury had the lowest melting point. It was the only metal to remain liquid at ordinary temperatures.

Since ancient times, chemists have discovered dozens of new metals, but none can shake the record low melting point of mercury. It was and remains champion. A number of the metals discovered in modern times, however, melt at the temperature of melting lead or less. Here in Table 36 on page 183, is the list:

There you are—the fourteen lowest-melting metals. Five of the eight lowest are the "alkali metals," which, in order of increasing atomic weight, are lithium, sodium, potassium, rubidium, and cesium. Notice that the melting points are 186, 97, 62, 38, and 28 respectively. The melting point goes down as the atomic weight goes up.

TABLE 36

Metal	Melting Point (° C.)
Mercury	−39
Cesium	+28
Gallium	30
Rubidium	38
Potassium	62
Sodium	97
Indium	156
Lithium	186
Tin	232
Bismuth	271
Thallium	302
Cadmium	321
Terbium	327
Lead	327

The melting point of cesium is second only to that of mercury (for stable metals anyway). A temperature of 28° C. is equivalent to 82.4° F. This means that cesium would be liquid at the height of a summer day's heat, and cesium is twice as common as mercury is. Could we play with cesium as we play with mercury if the day is hot enough?

Not likely. All the alkali metals are extremely active and react violently with, among other things, water. Let the alkali metals come in contact with the perspiration film on your hands and you will be sorry indeed. Since the alkali metals grow more active with increasing atomic weight, cesium is the worst of the lot that I've listed. No playing with cesium!

There is a sixth alkali metal, francium, with an atomic weight still higher than that of cesium. It is radioactive, has only been prepared in excessively minute quantities, and its chemical properties are not known. It would be safe to predict, however, that its melting point would be about 23° C. (73° F.) and it would therefore be liquid through most of a New York summer.

However, combine its chemical activity with its radioactivity, and the fact that only a few atoms at a time can be brought together—and forget francium.

Metals can be mixed to form alloys, and such mixed metals generally have a lower melting point than any of the pure metals making it up.

Suppose, for instance, we melt together 4 parts of bismuth, 2 parts of lead, 1 of tin, and 1 of cadmium, and let the mixture solidify. The result is "Wood's metal." While no component metal of the alloy melts at a temperature lower than 232° C., the alloy melts at 71° C. It is a "fusible alloy," one that melts below the boiling point of water. Lipowitz's alloy, in which the proportion of lead and tin is raised slightly, will melt at temperatures as low as 60° C.

Fusible alloys have their chief uses as safety plugs in boilers or automatic sprinklers. The recipe can be adjusted to give them a melting point slightly above the boiling point of water. A too-high rise in temperature melts them and allows steam to escape from the boiler, relieving dangerous pressure, or allows water to pass through the automatic sprinklers.

Fusible alloys are also used in practical jokes. A teaspoon made of Wood's metal is passed to someone who then innocently stirs his hot coffee while carrying on an animated conversation. To connoisseurs of such things, the look on the victim's face when he finds himself holding the mere stub of the handle of the spoon, is supposed to be delectable indeed.

You can also form alloys of the alkali metals which would melt at lower temperatures than any alkali metal alone—and which will, in some cases, melt at temperatures lower, even, than that of mercury.

But suppose we confine ourselves to solid metals we can handle with impunity. The alkali metals and their alloys cannot be touched. Neither can solid mercury, which is too cold for comfort. Let us ask what metals among those that *can* be touched are most easily melted.

There are the fusible alloys of which I've just spoken,

but lower melting than any of them is gallium—a pure metal, safe to touch, and melting at only 30° C.

And now that I have finally reached it, I intend to go on with its story—in the next chapter.

13 The Predicted Metal

I frequently receive letters from readers who attempt to find some insight into the mysteries of nature by shuffling facts or supposed facts into some kind of pattern. Very often, the readers are not professionals or experts in the subject they are trying to handle.

My own impulse, then, is to reject such attempts out of hand—but I never quite dare. I invariably ponder before answering, and even when I finally decide they are all wrong, I try to make my response a polite one. After all, one can never tell; and I have a peculiar horror against going down in scientific history as the man who laughed at So-and-So.

For instance, there is the man who laughed at John Alexander Reina Newlands, and I would gladly point the finger of scorn at him if I only knew his name.

Newlands was born in 1837 of an English father and an Italian mother, and he remembered his Italian ancestry well enough to fight with Garibaldi in 1860 for the unification of Italy. He was interested in both chemistry and music, and eventually became an industrial chemist specializing in sugar refining. In his spare moments, he turned his attention from sugar to the elements.

One had to wonder about the elements in those days. By 1864, about sixty different elements were known— elements of all kinds, sorts, and varieties. There seemed no logic to the list, however, no order. There seemed no way of predicting how many elements there might be altogether and for all anyone could say in 1864, the number might be infinite. Chemists were growing more and more depressed over the matter. If there were vast numbers of elements of all kinds, the universe would be unbearably complicated.

It is almost an article of faith with scientists that the universe is orderly and, basically, simple. Therefore, there had to be some way of finding order and simplicity in the list of elements. But how?

Newlands amused himself by juggling the elements in various ways. Over the past few decades, chemists had been carefully working out the atomic weights of the elements (that is, the relative masses of their respective atoms), and those figures now seemed hard and reasonably accurate. Why not, then, arrange the elements in the order of atomic weight?

Newlands did so, then listed them in a table seven elements in width. First, there were the seven elements of lowest atomic weight; then, under them, the next seven, and so on. It seemed to Newlands that certain groups of elements of very similar properties fell in vertical lines when this was done, and that this was significant.

Could it be that elemental properties repeated themselves in groups of seven? His musical interests led him irresistibly to remember that the notes of the scale repeated themselves in groups of seven. The eighth note ("octave") was almost a duplicate of the first. Musical notes repeated themselves in octaves, in other words. Might not the same be true of the elementss?

Newlands therefore wrote up his results in a paper presented for publication to the English Chemical Society and referred to his discovery as the "Law of Octaves."

The society rejected it with contempt, very much as

they might if I had submitted one of these speculative science essays of mine for similar publication. They had some reason for the rejection, though, for it had to be admitted that Newlands's table was quite imperfect. Although some very similar elements fell into columns, some extraordinarily *dis*similar elements did the same.

However, what really bothered the society, I am sure, was the whole notion of playing games with elements and making tabular arrangements. The very notion of listing the elements in order of atomic weight seemed a trivial trick, and one chemist (this is the wise guy I referred to at the start of the essay) asked Newlands why he didn't try to list the elements in alphabetical order and see what kind of a table he could squeeze out of that. I only hope the gentleman lived long enough to have to swallow his words. It would only have taken eleven years.

Actually, two years earlier (and quite unknown to Newlands) a French geologist with the formidable name of Alexandre Émile Beguyer de Chancourtois had also tried to list the elements in order of atomic weight. Instead of making a table, he imagined the list of elements wound helically about a cylinder. In this manner, he got much the same results Newlands had obtained in his table, but in nowhere nearly as simple a manner.

Beguyer wrote a paper on the subject and included a painstaking diagram of what his cylinder of elements would look like. The paper was published in 1862 but the complicated diagram was omitted. The omission of the diagram made the article impossible to follow. This was all the more so since Beguyer de Chancourtois was a poor writer who made free use of geological terms that were unfamiliar to chemists. His paper was completely ignored.

Despite the dangerous possibility of being laughed at, chemists continued, occasionally, to try to wring order out of the list of elements. As the 1860s closed, two men, independently, each made a try. One was a German named Julius Lothar Meyer and the other a Russian named Dmitri Ivanovich Mendeléev.

Matters had grown more subtle in the five years since

Newlands. Both the German and the Russian arranged elements in order of atomic weight, but both used other atomic properties as additional guides. Without going into detail here, I can say that Meyer made use of atomic volume and Mendeléev used valence.

Each man noted that when the elements were arranged in order of atomic weight, the other properties (such as atomic volume and valence) rose and fell in an orderly manner. They recognized further that the period of rise and fall did not always involve the same number of elements. At the start of the list the period was seven elements long, but later it grew longer. It was one of Newlands's errors that he tried to make the length of the period invariable, and that helped make it inevitable that dissimilar elements would fall into the same column.

Both Meyer and Mendeléev succeeded in publishing their work. Mendeléev managed to get into print first, publishing in 1869, while Meyer published in 1870. One might expect that Mendeléev, a Russian, would lose out even so because, in general, European chemists did not read Russian, and Russian discoveries were usually ignored in consequence. However, Mendeléev was foresighted enough to publish in German.

Even so, Mendeléev and Meyer might still have received joint credit, were it not for the difference in their approaches. Meyer was timid. Not at all eager to compromise his scientific career by stepping out too far ahead of the front lines, he advanced his conclusions in the form of a graph relating atomic volume to atomic weight. He made no attempt at interpretation, but let the graph speak for itself, which it did but softly.

Mendeléev, however, actually prepared a "periodic table of the elements" as Newlands had done, one in which the various properties varied in a periodic manner. Unlike Newlands, Mendeléev refused to allow any of the columns to contain unfitting elements. If an element seemed about to fall into a column which it didn't fit, Mendeléev moved it to the next column, leaving behind an empty hole.

How explain the empty hole? Mendeléev pointed out

boldly that it was quite obvious that not all elements had yet been discovered and the empty hole contained one of these undiscovered elements. Newlands had made no allowance for undiscovered elements. As for Meyer, his graph was so arranged that "holes" did not stand out, and Meyer himself later admitted he would never have had the courage to argue as Mendeléev had done.

From the properties of the remaining elements in the column with the empty hole, Mendeléev went on to insist he could even predict the properties of the undiscovered elements. He selected, in particular, the holes that were under the elements aluminum, boron, and silicon in his early tables. These holes, he said, contained undiscovered elements which he provisionally named "eka-aluminum," "eka-boron," and "eka-silicon" respectively.

(Eka is the Sanskrit word for "one" so that the name implies the "first element under aluminum" and so on. Since *dvi* is the Sanskrit word for "two," the two holes under manganese would be "eka-manganese" and "dvi-manganese" respectively. These are the only cases I know of where Sanskrit has been used in scientific terminology.)

Consider eka-aluminum, for instance. Judging from the rest of the column and from its position in the list generally, Mendeléev decided that its atomic weight would be about 68; that it would have a moderate density of 5.9 times that of water; that it would have a low melting point but a high boiling point; and that it would have a variety of carefully specified chemical properties.

The rest of the chemical world reacted to this with anything from a laugh of indulgent amusement to a snort of contempt. Playing with elements, and building complicated structures out of them, was bad enough, but describing elements one had never seen on the basis of such structures seemed like nothing more than mysticism—or even charlatanry.

I wonder if Mendeléev might not have been saved from worse criticism by the fact that he was Russian. Westerners must have felt indulgent toward the ravings of mystical Russians, and tolerated there what they would not have tolerated among their own countrymen.

But let's shift the focus to France again and to another Frenchman with a formidable name—Paul Émile Lecoq de Boisbaudran. He was a self-educated young man of means, fascinated with chemical analysis, and particularly attracted by the new technique of spectroscopic analysis in which heated minerals could be made to produce spectra composed of lines of differently colored light.

Each element produced its own spectral lines distinct to itself. This technique had been introduced in 1859 and its developers had almost at once found minerals yielding spectral lines not produced by any known elements. Orthodox chemical investigation into the minerals producing these lines demonstrated the existence of two new elements: cesium and rubidium.

Lecoq de Boisbaudran burned to discover elements, too, and moving into the field at the first announcement, he spent fifteen years subjecting every mineral he could get his hands on to spectroscopic analysis. He carefully considered the lines he obtained and moved intelligently in the direction of those minerals most apt to give him the new elements he wanted.

Finally he chanced upon a mineral which had been known to early mineralogists as *gadena inanis* or "useless lead ore." It was useless because it was a mixture of zinc sulfide and iron sulfide and procedures designed to get out the lead it didn't contain naturally failed. It is now called "sphalerite" from a Greek word meaning "treacherous" because it so often deceived the early miners.

This ore was anything but useless or treacherous to Lecoq de Boisbaudran, however. In February, 1874, he subjected the mineral to spectroscopic analysis and spotted two lines he had never seen before.

He hastened to Paris where he repeated his experiments before important chemists and established his priority. He then began working with larger quantities of the mineral and by November, 1875, worked it down to a gram of a new metal, enough to present some to the Academy of Sciences in Paris and to run chemical tests on the rest.

The new metal proved to have an atomic weight just under 70. It had a density of 5.94 times that of water, a low melting point of 30° C., a high boiling point of nearly 2000° C., and it showed a variety of specific chemical reactions.

This had no sooner been announced when Mendeléev, far away in Russia, pointed out excitedly that what Lecoq de Boisbaudran was describing was precisely the eka-aluminum that he had deduced from his periodic table five years before.

The chemical world was thunderstruck. Mendeléev's prediction of the properties of eka-aluminum existed in print. Lecoq de Boisbaudran's description of the properties of his new element existed in print. The two tallied almost exactly in every detail.*

There was no denying it. Mendeléev had to be right. The periodic table had to be a useful description of the order and simplicity behind the elements.

If there was any doubt, the other two elements described by Mendeléev were also found within a few years and predictions were again found to tally with the fact. As all the ridicule had fallen on Mendeléev before and none on Meyer, so now all the credit went to Mendeléev. In 1906, just a few months before his death, Mendeléev almost received the Nobel Prize, losing out by just one vote to Moissan, the discoverer of fluorine.

Both Newlands and Beguyer de Chancourtois lived to see themselves vindicated. After Beguyer de Chancourtois died in 1886, a French journal, in remorse, published his diagram of the cylinder, the one that had not been published thirty years before. And in 1887, the Royal Society

---

* Lecoq de Boisbaudran first got a figure of 4.7 for the density, but Mendeléev insisted that that could not be so. And he was right. The Frenchman's first samples were too impure. After proper purification, the density figure matched the prediction. A discrepancy in which the prediction won out over the first observation made the situation even more dramatic.

finally gave Newlands a medal for the paper which the Chemical Society had refused to publish.

As for the element discovered by Lecoq de Boisbaudran, he took the privilege of the discoverer and named it. Out of noble patriotism, he named it for his native land. For this he turned to ancient Rome and used the Latin name Gallia ("Gaul" in English), so that the new element became "gallium."

Yet was it patriotism, pure and unalloyed? *Lecoq* in English is "the cock," and that noble barnyard fowl is *gallus* in Latin. Well, then, was gallium named for Gallia, the nation, or for *gallus*, the discoverer himself? Who can say?

Gallium is not an extremely common element, but on the other hand, it isn't a particularly uncommon one, either. It is about as common as lead, which makes it about thirty times as common as mercury and three thousand times as common as gold. Unfortunately, gallium is much more evenly spread out through the earth's crust than any of these other elements, so that there are few places indeed where it can be found in sufficient concentration to make its extraction practical.

Its richest source is a mine in Southwest Africa and even there the ore is only about 0.8 per cent gallium (as compared, though, with a usual figure of 0.01 per cent). Several thousand pounds of metal are all that is produced in the course of a year.

The most spectacular property of gallium is its melting point, which is 29.75° C. (or 85.55° F.). This means that it is solid usually, but will melt on a hot summer day. Notice, too, that its melting point is below the normal temperature of the human body (37° C. or 98.6° F.). This makes it possible to perform a dramatic experiment.

You begin with a rod of solid gallium. That's a moderately expensive proposition right there, for the present cost of gallium is something like $50 an ounce. Oh well, you *borrow* a rod of solid gallium.

Naturally, you choose a day when the temperature is in the seventies and you hold the rod in a pair of cool tongs.

Now bring the end of the gallium rod down into the palm of your hand and leave it there.

Slowly, the warmth of your hand suffices to melt the gallium. The rod shrinks, and gathering in your palm will be a puddle of a silvery liquid that looks like mercury. No other pure metal will produce this effect and it is quite creepy to watch something that looks like steel puddle up in that manner.

Liquid gallium is perfectly safe to hold, but is less than half as dense as mercury so it doesn't have the latter element's unusual weight. Furthermore, liquid gallium wets glass as mercury does not, which deprives the former of some interesting effects.

The melting point of gallium can be lowered still further if it is mixed with other comparatively low-melting metals. With the proper admixture of indium and tin, an alloy with a melting point as low as 10.8° C. (51.4° F.) can be produced. It can replace mercury for frictionless electric contacts with moving parts.

Once gallium melts, it freezes again only with reluctance, even if it is cooled quite a bit. It will stay liquid even in an ice bath. However, if a piece of solid gallium is dropped into such "super-cooled" liquid gallium, the solid acts as a "seed" about which the gallium atoms line up in orderly fashion, so that the liquid solidifies at once.

In the case of most substances, the liquid form is less dense and more voluminous than the solid form. A liquid therefore generally shrinks when it solidifies. There are several exceptions to this, however, and the most important of them is water.

Where liquid water has a density of 1.00 gram per cubic centimeter, ice has a density of 0.92 gram per cubic centimeter. This means that 10 cubic inches of liquid water will freeze to form 11 cubic inches of ice. The pressure required to shrink that ice back to 10 cubic inches is enormous. If water is allowed to freeze in a closely fitting sealed container, that enormous pressure must be exerted on the ice to keep it from expanding. There are few containers that can do so and even strong ones will shatter.

Gallium is another exception. The density of the liquid is 6.1 grams per cubic centimeter, while that of the solid is 5.9. This means that 29 cubic inches of liquid gallium will freeze to 30 cubic inches of solid. The pressures developed are not as enormous as in the case of water, but they are large enough. For that reason, gallium is shipped in flexible containers of rubber or plastic. If, in the course of shipping, the metal melts and resolidifies, the containers bulge and distort rather than break.

Gallium is remarkable not only for its low melting point, but also for its high boiling point. It boils at 1983° C. so that at ordinary temperatures, or even at a red heat, it produces insignificant quantities of vapor. In this it is quite different from mercury, which boils at 357° C. and which produces sizable amounts of vapor even at ordinary temperatures. Mercury vapor is toxic which means that the metal should be handled with considerable care. Gallium, on the other hand, poses no problems of that sort at all.

This combination of low melting point and high boiling point on the part of gallium raises a point.

Mercury is used in thermometers because it is liquid throughout the ordinary range of temperatures in which chemists are interested. To record temperatures below -39° C. chemists have to use alcohol thermometers, for ethyl alcohol doesn't freeze till a temperature of -117° C. is reached. For temperatures lower still, other dodges are used.

What about temperatures in the ranges higher than a mercury thermometer can handle? For that we need some substance that will not change with heat (an element is safer than a compound in this respect), that will be liquid through part of the liquid range of mercury and that will remain liquid at temperatures as high as possible.

In other words, what elements have melting points lower than 357° C. and boiling points higher than 357° C.? There are exactly fourteen of them and I am going to arrange them in Table 37 in order of decreasing length of temperature range over which they remain liquid:

TABLE 37

Element	Melting Point (° C.)	Boiling Point (° C.)	Liquid Range (degrees)
Tin	232	2270	2038
Gallium	30	1983	1953
Indium	156	2000	1844
Lead	327	1620	1293
Bismuth	271	1560	1289
Thallium	302	1457	1155
Lithium	186	1336	1150
Sodium	98	880	782
Potassium	62	760	698
Rubidium	38	700	662
Cesium	28	670	642
Selenium	217	688	471
Cadmium	321	767	446
Sulfur	113	445	332

As you see, three of the elements listed have a liquid range nearly or quite 2000 degrees. This is very unusual, for there are no others like it in the entire list of elements. (There are, to be sure, a number of elements that remain liquid for 2000 degrees and more, but that have their liquid range in a very high and inconvenient spread of temperature. Osmium, for instance, remains liquid for 2600 degrees from 2700° C. to 5300° C. That's pretty useless.)

For the "high, but useful" range, which would take us up to 2000° C., we have just the three elements: tin, gallium, and indium in that order.

As it happens, none are elements that are available in large quantity, but then large quantities are not needed for thermometers. Of the three, gallium is the lowest melting and therefore far the easiest to handle in a technique that must use liquid in filling the thermometer.

For that reason, gallium thermometers have indeed been used. As the fine column of mercury is enclosed by glass in the ordinary thermometer, so the fine column of liquid gallium is enclosed by quartz in the gallium thermometer. Such gallium thermometers are particularly useful in the range from 600° to 1500° C.

Gallium is proving to be useful in all sorts of solid-state devices, for which purpose it must be prepared (and is) with impurity concentrations of not more than one part in a million.

One of its compounds, gallium arsenide (GaAs), can be used in solar cells, converting sunlight directly into electrical current. It can be used as a semiconductor and in transistors at temperatures beyond those at which more ordinary devices of that sort will work. There is every reason to think it can be used to produce a laser beam.

There is no doubt but that gallium can make a satisfactory splash in the new world of far-out science, but nothing that happens to it is likely to outweigh the glamour and significance of the manner of its discovery.

C BIOLOGY

14 The Terrible Lizards

James D. Watson has recently published a book, *The Double Helix*, in which he details the inside story of the discovery of the structure of the DNA molecule. The book is making the headlines not because of the importance of the subject but because of the fact that it exposes scientists as human beings possessed of human failings.

Well, why not? A great intellect need not necessarily be accompanied by a great soul. There are villains among scientists as among any other group.

My own pet candidate for a high place in scientific villainy is Sir Richard Owens, a nineteenth-century English zoologist. He was the last of the top-rank "nature philosophers" who had taken up the mystical notion of the German naturalist Lorenz Oken. These believed in evolutionary development through vague internal forces that drove creatures on to some particular end.

When, in 1859, Charles Darwin published *The Origin of Species*, in which he presented evidence for evolution by natural selection, Owen was horrified. Natural selection, as Darwin described it, was a blind force, changing species

198

through its action on random variations among individuals.

Owen could not accept evolution by random effects and he came out against Darwin. That, of course, was his right. It was even his scientific duty to disagree with all his might. Darwin's suggestion, like all scientific suggestions, had to survive the battles fought in the intellectual arena, and no honorable weapon was outlawed in such battles.

No *honorable* weapon. Owen chose to review Darwin's *Origin of Species* in as many different outlets as he could wangle. He chose to make those reviews anonymously and to quote extensively and with worshipful approval from his own work, making himself sound like a crowd. He chose to give an unfair summary of the contents of the book and to ridicule it rather than to present opposing testimony objectively. Worst of all, he urged others to denounce Darwin, vitriolically and unscientifically, before lay audiences, feeding them the necessary misinformation for the purpose.

In short, Owen was cowardly, spiteful, and contemptible, and it is a source of gratification to me that he lost out.

Yet that can't be allowed to obscure the fact that he made important contributions to biology. He discovered the parathyroid glands in 1852, while dissecting a rhinoceros. (It was to be quite some years before they were found in man as well.) He was the first to describe the recently extinct moas of New Zealand and the anything but extinct (alas!) parasite that was later found to be the cause of trichinosis.

His greatest fame to the general public, however, lies in a single word. He was one of the earliest of those who studied the fossils of certain giant creatures, long since extinct, which soon caught the imagination of the world. Weighing up to five times the size of the largest living elephant, they shook the ground between 70,000,000 and 270,000,000 years ago.

The huge skeletons that were built up out of the fossilized remnants were clearly reptilian in nature, so Owen called them "terrible lizards," and since he used Greek, that became "Dinosauria." (Actually, those ancient giant reptiles are more closely related to alligators than to lizards, but I must admit that "Dinocrocodilia" would have been a rotten name.)

The name caught on, and today, I am sure, many American children can describe various dinosaurs even though they can't describe a hippopotamus and never even heard of an okapi.

Yet for all the fame of the word, and for all its continuing popularity, it has passed out of the scientific picture. As it turns out, there is no single group of animals we can call by that name. The term is gone from the chart of animal classification. You can look it over from top to bottom and you will find nothing labeled "Dinosauria." (Ha, ha, for you, Sir Richard Owen.)

Furthermore, dinosaurs are not necessarily large and monstrous. Many of them were quite small and were considerably less "terrible" than, say, an angry police dog. Then, too, some of the large extinct reptiles that look terrible indeed are *not* considered dinosaurs in the strictest sense of the word.

So let's go into the matter of the terrible lizards and find out what they are and what they are not.

In classifying ancient reptiles one must, of necessity, use bone structure as the source of distinctions, for it is only the bones of these long-dead creatures that we have left to study. The skull is used quite often, because it has a complicated structure of many bones, and this structure appears in convenient broad variations.

For instance, the class "Reptilia" is divided into six subclasses in accordance, to a large extent, with the structure of the skull. The most primitive skulls have the bony structure behind the eye socket solidly enclosed. Such a skull belongs to the subclass Anapsida ("no opening").*

---

* Actually, the Greek *apsis* means "wheel," "arch," "vault," and a few other things. If we take the meaning of "wheel," and consider that it might be applied to a roughly circular opening, we might as well translate the word as "opening" and make it clear. In this article, I will try to give the literal meanings of the zoological terms, which are almost always in Greek or

The earliest important reptiles, of the order Cotylosauria ("cup-lizards," because of their cup-shaped vertebrae), had anapsid skulls. These cotylosaurs, low-slung, stocky, not more than six feet long, and not terribly far advanced over their amphibian forebears, existed about 300,000,000 years ago. They are often called the "stem-reptiles" since they represent the stem of the reptile family tree, from which all later forms branched off—though the cotylosaurs themselves are long since extinct.

One group of their descendants, who appeared early (perhaps 230,000,000 years ago), and only one, retained the anapsid skull. Oddly enough, this primitive order still exists, although more advanced cousins have long since grown extinct. This is the order Chelonia ("turtle"), which, of course, includes the turtles and tortoises.

Another type of reptilian skull has an opening behind the eye socket. Creatures with such skulls make up the suborder Synapsida ("with opening"). This entire suborder is extinct; at least, there are no living reptiles with synapsid skulls. However, descended from these Synapsida are the mammals of today.

There are two other kinds of reptilian skulls with a single opening behind the eye sockets. The arrangement of the bones around that opening differ in the two cases, and both differ from the arrangement in the synapsid skulls. So we have the suborders Parasida ("side opening") and Euryapsida ("wide opening"). Both of these suborders are totally extinct. There are no living reptiles with either paraptid or euryaptid skulls. Nor have any non-reptilian forms been descended from them.

The most familiar of the parapsids are the Ichthyosauria ("fish-lizards"). This is a good name because at their first known appearance, 220,000,000 years ago or so, they had already been living in the sea so long as to become completely adapted to it. They had assumed the streamlined fish-shape just as some of our modern sea mammals

---

Latin, of course. I must not conceal from you that I don't always know why those literal meanings have been chosen. If any Gentle Reader knows, I will welcome the information.

have. In fact, they very closely resemble long-snouted dolphins.

Of course, when we speak of ichthyosaurs, we don't speak of a particular animal, but of a large group of diverse animals. There were species of ichthyosaurs that were no more than 2 feet long, for instance, and other species which reached lengths of 60 feet. The largest ichthyosaurs were the size of the modern sperm whale and were, in their time (180,000,000 years ago), the largest animals alive. By 40,000,000 years later, the giant species had died out and considerably smaller species, with shorter tails and no teeth, took their place.

We arrive now at a tricky point. The ichthyosaurs, despite their outward fishlike appearance, are complete and thoroughgoing reptiles. They are quite extinct and some of them were enormous. Does not this qualify them— as large, extinct reptiles—to be considered among the dinosaurs?

In popular speech, they surely are, but to purists this is not proper. Almost all the creatures popularly called dinosaurs belong to a particular subclass of Reptilia, and Parapsida is not it. Strictly speaking, creatures outside that one subclass have no claim on the term and in that sense the ichthyosaurs are not dinosaurs.

Passing on to Euryapsida, we have as the best-known examples other aquatic creatures only slightly less well adapted for sea life than the ichthyosaurs. All have limbs adapted for paddling and swimming. Some look as though they can still hobble around on land, but one group is so completely paddle-equipped that it clearly cannot leave the sea. These are the Plesiosauria ("near-lizards," because they looked pretty much like reptiles except for the oddity of four paddles in place of ordinary limbs).

If the ichthyosaurs were the reptilian equivalent of whales and dolphins, the plesiosaurs seem to be reptilian seals. Their most remarkable attribute, perhaps, were the long necks possessed by most (but not all) of them. These were probably sent darting forward, like animated spears, after fish. At the height of the age of reptiles, 100,000,000 years ago, certain varieties were 50 feet long or more, with necks making up two-thirds of this total length. One of these gigantic varieties, the Elasmosaurus ("plated lizard"), probably had the longest neck ever existing on Earth.

The plesiosaurs look much more like the ordinary notion of the dinosaur than the ichthyosaurs do. They have small heads, long necks and tails, and a barrel-like body. And yet the plesiosaurs are not dinosaurs because they, too, like the ichthyosaurs, belong to the wrong subclass.

That brings us to a fifth kind of reptilian skull. In a purely reptilian sense, that fifth is the most successful by far. (I qualify the statement because the synapsids, although only moderately successful as reptiles, did give rise to the mammals, and that must be considered an enormous, if non-reptilian, success.)

In this last kind of skull, there are two openings behind the eye socket; such a skull is a "diapsid" ("two openings") skull. There is, however, no one subclass bearing that name. The reason for that is that there are no less than two important groups of reptiles with diapsid skulls and neither can claim sole possession of the title.

The first of the diapsid subclasses is Lepidosauria ("scaled lizards," from Greeks). This includes the order, Squamata ("scaly," from Latin), which in turn includes the most successful of living reptiles, the snakes and lizards.

Another order of lepidosaurs, Rhynchocephalia ("snout-heads," because they have prominent, beaky snouts), is interesting for a completely different reason; not because it is flourishing, but because it has avoided extinction by the narrowest possible margin—a single rare species. The order was never very important and, except for that one survivor, died out about 70,000,000 years ago. But there remains that one survivor. It is a moderately large lizard-like creature, no more than 30 inches long, at most, from snout to tail tip. In recent times it was still to be found on the main islands of New Zealand, but no longer. It is now to be found only on a few offshore New Zealand islets, where it is sternly protected by law. Its common name is tuatara ("back-spine," in native Maori, since in addition to the scales that cover its body, it has a line of spines down its backbone). Its more formal name is sphenodon ("wedge-tooth"), which is the name of its genus.

Despite its looks, it is not a lizard. It has a bony arch in its skull that no lizard possesses (the first giveaway to the early dissectors that they had something unusual in hand). Their teeth are attached in non-lizard-like fashion and it also has a "nictitating membrane" in its eyes, which birds possess, but not lizards. Finally, it has a particularly well-developed pineal gland at the top of its brain, something lizards do not have nearly as well-developed. In the young sphenodon, it bears the anatomical appearance of a third eye, though there is no indication yet that it is light-sensitive.

The second diapsid subclass is Archosauria ("ruling reptiles"), and it is this subclass, and this subclass only, as the name implies, to which dinosaurs belong.

Of course, we might stop to ask why two subclasses are made out of creatures that all have diapsid skulls. Well, there are other distinctions. For one thing, the archosaurs have teeth set in sockets and the lepidosaurs (with one very minor exception) don't. This seems an unimportant distinction to the layman, perhaps, but it isn't. The improvement in tooth efficiency is such that the archosaurs became, for a time, the most successful of all the reptilian subclasses. Besides, one such distinction is usually representative of a whole family of distinctions.

The most primitive archosaurs make up the order Thecodontia ("socket teeth"), and it is from these that all other archosaurs developed, though they themselves are long since extinct.

Many of the thecodonts were rather on the small side and adopted a bipedal posture. The forelimbs were reduced in size, the hind limbs enlarged and strengthened, and a long balancing tail was developed. They rather looked like reptilian kangaroos.

Certain heavier and clumsier thecodonts, however, were forced to remain on all fours, and these developed into the order Crocodilia, which survives, of course, to this day. The alligators and crocodiles are the only living reptiles that belong to the subclass Archosauria, the one to which the dinosaurs belonged. And yet crocodiles are not to be considered dinosaurs, even though they are their closest living relatives. The dinosaurs are restricted to two orders, and only two orders, within the subclass. The alligators and crocodiles are outside those two orders and are therefore not dinosaurs.

More spectacular is another group of thecodont descendants, which make up the order Pterosauria ("winged lizards"). These were lighter still, developed long thin webs based on enlarged little fingers, made wings out of them, and were the only reptiles ever to engage in true flight.

The early pterosaurs had long heads with sharp teeth, and long tails, too. The later ones grew larger, had much shorter tails, and often lacked teeth altogether. About 150,000,000 years ago, the skies held the largest of all the pterosaurs. This was Pteranodon. It had a wing span of about 20 feet and a long-crested skull some 3 feet from end to end.

But the pterosaurs were not the only descendants of the thecodonts to learn the secret of true flight. Another group converted its scales into feathers. From these descended a group of creatures, the birds, so radically different from other reptiles in so many ways, as to deserve being placed in a class of its own, "Aves."

There now remain the two orders of archosaurs (both entirely extinct) that contain the dinosaurs. To simplify their relationship to other reptiles, I have prepared Figure 8, which deals mainly with zoological classification and is *not* primarily concerned with lines of evolutionary development.

If the two orders could be grouped into one, that combined order would undoubtedly have retained Owen's title of "Dinosauria." However, a closer study of the creatures early showed that there were important distinctions to be made among them, notably in the pelvic girdle (or hipbone).

FIGURE 8 THE REPTILES

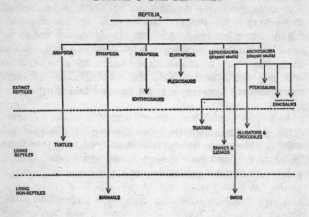

Since the ancestral dinosaurs had the bipedal gait of the thecodonts, the pelvic girdle had to bear the full weight of the animal. The pelvic girdle was strengthened therefore and its three chief bones enlarged in the course of evolution. The uppermost bone of the girdle, the ilium, was enlarged and came to be fused to the backbone to produce a structure of great solidity and strength. This fusion is characteristic of both orders and is therefore an important characteristic mark of the dinosaur.

There are two other bones to the girdle, however. In some dinosaurs these remained well separated and were set nearly at right angles. This rather resembles the situation in living lizards and so all dinosaurs with this arrangement of hipbones are placed in the order Saurischia ("lizard-hip").

In the remaining dinosaurs, the two lower bones of the pelvic girdle are lined up closely parallel and slant backward. This arrangement resembles that in birds and so these are placed in the order Ornithischia ("bird-hip").

Nor is the distinction a petty one in zoological terms. So noticeable is it, that once the difference in hip girdles is explained, anyone at all can tell whether a dinosaur belongs to one group or the other by one quick glance at the skeleton.

It is for this reason that "dinosaur" is no longer a formal zoological term. One can, and often does, speak of "saurischian dinosaurs" and "ornithischian dinosaurs." It is more appropriate, however, to speak of "saurischians" and "ornithischians," leaving out the word "dinosaur" altogether.

The saurischians had their heyday first. These are divided into two suborders, Theropoda ("beast-feet") and Sauropoda ("lizard-feet"), because the toe bones of the former more closely resemble those of mammals in number than do the toe bones of the latter.

An easier way of distinguishing the two suborders is to remember that the theropods are bipeds and the sauropods are quadrupeds.

The earliest of the theropods were, in fact, very much like the thecodonts—light, small bipeds adapted for fast running. These were the Coelurosaurs ("hollow lizards," because their bones were hollow for the sake of lightening the body structure). Many of these were quite small, and one, Compsognathus ("elegant jaw" because it was so small and delicate), was only about the size of a chicken and was the smallest of the known dinosaurs.

However, there was a general tendency for species to grow larger as the ages wore on—perhaps because increasing competition among the various dinosaurs put an increasing premium on strength. By the late Cretaceous, 80,000,000 years ago, coelurosaurs the size of ostriches had evolved. One of them was almost exactly the size and shape of an ostrich, with a small head sporting a horny, toothless beak, a long neck, and powerful legs. Even though it had forearms with clutching fingers instead of stubby, vestigial wings, and a long tail instead of plumes, it is still called "Ornithomimus" ("bird-imitator").

Another line of theropods was the "carnosaurs" ("meat-lizards"), so called because they were characteristically meat-eating. So were the coelurosaurs, as a matter of fact, but the physical appearance of the carnosaurs made the fact much more horribly evident in their case.

The carnosaurs retained the bipedal structure but went in for size far beyond anything the coelurosaurs could do. By the late Cretaceous, this had reached its maximum in Tyrannosaurus ("master-lizard"), whose four-foot-long head was carried some nineteen feet above the ground. The full length of its body, from snout to tail tip, was probably something like 50 feet, but its forelegs were tiny, not much longer than a man's, and far too short to be of any use. They couldn't even reach the mouth.

The jaws of the Tyrannosaurus could do their work without help, however. Its many teeth were up to six inches long and it is clear from its skeleton alone that it was the most nightmarish creature that ever clumped the Earth. It is the largest land carnivore on record, being at least as massive as the largest elephant (which is herbivorous) that ever lived.

The enormous thighs of the Tyrannosaurus show clearly that it was approaching the practical limits for bipedality.

The sauropods were gigantic creatures, and the most familiar, in appearance, of all dinosaurs. They were super-elephantine in structure, with long necks at one end and long tails at the other. Indeed, they looked like giant snakes that have swallowed giant elephants, with the columnar legs of the latter breaking through and walking off with the creature.

There are clear signs of the bipedal ancestry of the sauropods, for all that they clumped down so hard on four legs. In most cases, the forelegs remained shorter than the hind legs so that their backs sloped upward and reached a peak at the hips.

The longest of all the sauropods is Diplodocus ("double beam"). Some specimens have been found which measured nearly 90 feet from the snout to the tip of its long tapering tail. No other animal was ever longer except for some of the very largest whales.

However, the diplodocus was a slenderly built creature and was by no means the most massive dinosaur. The

Brontosaurus ("thunder-lizard"), though shorter, was more massive and may have weighed up to 35 tons.

More massive still was the Brachiosaurus ("arm-lizards," so called because, in the course of evolution, its forelimbs had finally developed to the point where they had overtaken the hind limbs and were longer). The Brachiosaurus may have weighed up to 50 tons and was the largest land animal that ever lived.

It is hard, though, to be sure how far we can justify the phrase "land animal." It is very likely that the large sauropods though they could clump about on land, if necessary, lived chiefly in rivers and lakes as modern hippopotami do, and for the same reason. They found their food there, as well as a certain protection, and the water buoyed up their giant weights.

The ornithischians, which were the more specialized of the two groups, did not come into their own until about 150,000,000 years ago, tens of millions of years after the saurischians had already developed into a variety of flourishing forms.

The ornithischians were all herbivores and their smaller representatives also retained the bipedality of the original dinosaurian ancestor, though their forelimbs never got as small as was the case among the saurischian bipeds.

Typical were the duck-billed dinosaurs, which developed a broad, flat jaw to handle their vegetable diet. The largest of these, Anatosaurus ("duck-lizard"), stood 18 feet high. It might resemble a Tyrannosaurus when seen quickly, from a distance, but it was quite harmless unless it stepped on you or fell on you.

Most of the ornithischians developed protection against the carnosaurs by developing armor of one sort or another. One of the best known is Stegosaurus ("roof-lizard"). It received its name because its skeleton was found in association with large bony plates which, it was first assumed, protected its back like shingles on a roof. Closer study showed that they stood on end in a double row from neck to the root of the tail, while the tip of the tail was armed with two pairs of long, pointed bony spikes.

The stegosaurus showed clear signs of ancestral bipedality, for its front legs were little more than half the length of the hind. Its tiny head contained a brain no larger than a modern-kitten's, though it was 30 feet long and more massive than an elephant. The stegosaurus is the very epitome of dinosaurian brainlessness.

It became extinct in the early Cretaceous, probably before Tyrannosaurus appeared on the scene. The famous sequence in Walt Disney's production *Fantasia* in which a Tyrannosaurus attacks and kills a Stegosaurus, though highly effective, is very likely anachronistic.

An armored ornithischian that evolved later than Stegosaurus and was indeed contemporary with Tyrannosaurus was Ankylosaurus ("crooked lizard"), and this was probably the most heavily armored creature of all time. It was a low, broad dinosaur that could not be easily overturned to expose its unarmored belly. Its back, from skull to tail, was layered with massive bony plates, which, along the sides, were drawn out into strong spikes. The tail ended in a bony knob that probably had the force of a battering ram when swung. It was a veritable living tank and I wonder what a battle between *it* and a Tyrannosaurus would have been like.

Finally there is the Triceratops ("three-horned"), which was built like a super-rhinoceros and is the best-known of a large and varied family. Its armor was concentrated in its head region. A broad frill of bone, six feet across, extended from the head and covered the neck. The face bore three horns, two long sharp ones over the eyes and a shorter, blunter one on the nose. In addition, the mouth was equipped with a strong, parrot-like beak.

For a summary of dinosaur relationships, see Figure 9.

But then came the end of the Cretaceous, 70,000,000 years ago, and something happened; we don't know what. All the dinosaurs that then existed, both saurischian and ornithischian, died off in a relatively short time, say a couple of million years. So did the spectacular non-dinosaur reptiles, the ichthyosaurs, plesiosaurs, and pterosaurs. So did some spectacular creatures who were not reptiles, such as the invertebrate ammonites.

FIGURE 9 THE DINOSAURS

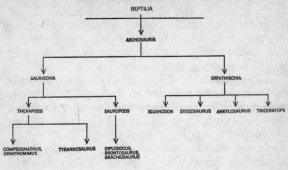

There have been almost as many theories to account for this as there have been paleontologists, and lately there was published a particularly interesting one which I will discuss in the next chapter.

15 The Dying Lizards

Nearly twenty years ago I wrote a story called "Day of the Hunters" in which, in fictional form, I presented a new theory to account for the sudden death of the dinosaurs at the end of the Cretaceous period, 70,000,000 years ago.

The theory was a simple one. Toward the end of the Cretaceous, I suggested, a certain group of small dinosaurs had developed intelligence, invented missile weapons, and hunted all the other dinosaurs to extinction. Then, for lack of any other prey, they hunted themselves to death, too.

Why are there no records of intelligent dinosaurs, then, or say, dinosaurs with large brain capacities? Well, intelligent creatures leave few fossils. Look how few primate fossils we discover, and primates are much more recent than the dinosaurs. As for artifacts . . .

But I am not here to defend my thesis, which, actually, I don't think is defensible. I used it merely to write a little story (which turned out to be what the science-fiction critic, Damon Knight, would call "minor Asimov") that would develop a not-too-subtle moral for our own times.

The problem remains, though. What killed off the dinosaurs?

For 150,000,000 years, an astonishing series of bulky reptilian species had dominated Earth's life forms. (I will call them "dinosaurs" in this essay even though, as I explained in the preceding chapter, this is an inadequate term.) All through this 150,000,000-year period, from 220,000,000 years ago to 70,000,000 years ago, individual species of dinosaurs became extinct, sometimes without leaving descendants as far as we know, and sometimes having previously branched off other species which, in a sense, replaced them. On other occasions, a species might grow extinct in the sense that it underwent slow changes that turned it into a new species, or into several new species.

About 70,000,000 years ago, however, quite suddenly (say, within a couple of million years perhaps) all the remaining dinosaurian species became extinct, leaving no descendants behind.

A hundred fifty years ago this was easy to explain, because at that time the doctrine of "catastrophism" was popular among biologists. In an age when the Bible was still revered as literal truth, biologists had to square the gathering evidence in favor of an Earth, and of fossil creatures, both many millions of years old, with a Biblical tale that made creation of both Earth and life seem to have taken place merely 6000 years ago.

A hint to the solution was found in the tale of the Flood. A Swiss naturalist, Charles Bonnet, suggested in 1770 that fossils were remnants of extinct species that had been destroyed in any of a series of world-wide cataclysms of which the Biblical Flood was only the most recent. After each such cataclysm life would begin anew, and Biblical truth could be preserved by saying that it dealt only with the most recent of several different creations.

The most prominent exponent of such catastrophism was the French naturalist Georges Cuvier, who, in the first decades of the nineteenth century, was the foremost student of fossils. With great skill he compared the anatomy of the fossils and showed that they could be arranged in a logical manner and fitted into still existing phyla. Using hindsight, we can see that what he did fairly shouted "Evolution!" at the top of its voice.

But Cuvier did not accept evolutionary explanations. Instead, he carefully pinpointed the location of four places in the fossil record where there seemed gaps. These, he held, were four examples of Bonnet's catastrophes.

Alas for Cuvier! More and more fossils were discovered and their order in time was more and more clearly worked out. And all the gaps disappeared. There is no point in time from the moment the fossil record begins (at a point we now know to be 600,000,000 years ago) to the present, where all forms of life cease to exist. Life was created only once.

In fact, there are species still alive and flourishing today that have existed with little change ever since before the time of the dinosaurs. The horseshoe crab is an example: it has not changed very much in 300,000,000 years.

Yet there have been times in the history of life when a great many species have indeed "suddenly" ceased to exist, while a great many other species kept right on going, and this is hard to explain.

A "partial catastrophe" must have taken place 70,000,-000 years ago. Something happened that killed off many species in a wide variety of habitats—the pterosaurs in the air and the ichthyosaurs in the sea, as well as the clumping land dinosaurs, while leaving other species intact. The early ancestors of the birds and of the mammals lived right through the end of the Cretaceous. So did the ancestors of the reptiles that still live today—even the ancestral crocodiles, which were not-too-distant relatives of the dinosaurs. And plant life lasted through the end-of-the-Cretaceous dividing line practically untouched.

What the answer might be nobody knows, but there have been a number of interesting speculations on the matter.

For instance, there might have been a climatic change. The dinosaurs may have been adapted to a mild Earth of flat land and shallow seas, with little seasonal variations. Then came a period of mountain building. The land heightened and grew rugged; the sea deepened and grew cold; the seasons became more extreme—and the dinosaurs died off.

I don't like this myself—at least, not as a sole explanation. Surely the Earth didn't get climatically unsuitable everywhere. Creatures have managed to hang on to restricted habitats when things went bad. The giant redwoods cling to places in California; the tuatara clings to its islands off New Zealand. Surely, there must have remained mild and marshy areas where at least some of the smaller dinosaurs might have hung on, at least for a while.

And could climatic changes alone kill the ichthyosaurs in the relatively unchanging environment of the sea?

Or perhaps it was the living environment that did it. The little furry proto-mammals, scurrying through the underbrush and doing their best to evade the eyes of the lordly reptilian masters, might nevertheless have grown fat on dinosaurian eggs left to care for themselves by the dim-witted saurian parents.

And eventually the mammals might have eaten enough eggs to block the generations and awoke one morning to find the reptiles gone. It is a dramatic story in a way and it suits us right down to the ground since it presents us mammals as heroes (if skulking egg-eaters can be considered heroes).

There are difficulties, of course. Primitive mammals had been in existence for a hundred million years by the time the Cretaceous period drew near its end. We must suppose that they suddenly increased in numbers and began to take an unbearable toll of dinosaur eggs. Or else we can decide that certain new species developed that specialized in these eggs, while leaving reasonably untouched the eggs of the ancestral crocodiles, lizards, snakes, and turtles.

And, for that matter, how did these mammals get at the eggs of the ichthyosaurs, which brought forth their young alive—and in the sea, in any case.

Then there are the falling-domino modifications that make capital of the fact that life is interdependent. Why should we suppose something had happened that affected every single one of the extinct species alike? Perhaps only a relatively small number of species were affected, and when these began to dwindle and die out, other species which depended upon the first set for food or for other necessities also died, and these in turn brought about the extinction of others—until a whole swatch of the fabric of life was cut out of existence.

This must happen all the time. It can easily be seen as a threat now. If the eucalyptus tree were to become extinct, the koala would have to become extinct, too, for it will eat nothing but eucalyptus leaves. If the zebra population were to vanish overnight, the African lions would drastically decrease in numbers. It doesn't even have to be a matter of food. Wipe out bees and numerous species of plants that depend on bees for cross-pollination will be wiped out also.

Something like that may have happened at the end of the Cretaceous. A group of species that formed part of a particularly tight interweaving of life died out, and with them the rest of the web went.

But what could the initiating factor have been?

Could a climatic change have killed some species and set the dominoes to falling? Did a group of egg-eating mammals kill off *some* species? Was it perhaps the advent of some new strain of bacteria or virus that killed off certain species in a vast plague?

Was it, on the other hand (as I have seen suggested), a plant evolution? Did the development of some precursor of modern grasses, which are hard, tough, and ruinous even to the highly adapted molars of the modern horse, bring about the end? The herbivorous dinosaurs, used to softer, more succulent vegetation (and possessing teeth to suit) began, perhaps, to decline as the grasslike plants spread more and more at the expense of the earlier species. And with the herbivores dying, the carnivores that fed upon them had to starve as well.

It remains only to choose the particular mechanism that set the dominoes to falling, and so far no one has been able to. There are too many possibilities to choose from and no reasonable evidence upon which to base the choice.

Indeed, I haven't even discussed all the possibilities yet. So far, I've mentioned only causes that could be one-shots or, if periodic, totally unpredictable. After all, when will there be another really radical weather change? When will a new plague come? When will there be the equivalent of creatures to eat our eggs or plants to set our cattles' teeth on edge?

It is much more interesting, in a grisly sort of way, to speculate on the possibility of reasonably predictable periodic occasions on which there would be a Great Dying. We do, in fact, find signs in the fossil record of periodic events of this kind, with the one at the end of the Cretaceous the most spectacular only because it is one of the most recent and therefore has its fossil record best preserved. There was a still more recent Great Dying of the huge mammals only a couple of million years ago. (By speculating on such periodic Great Dyings, be it noted, we turn the scientific wheel full cycle and are back to something a little bit like Cuvier's catastrophism. This often happens in science.)

Let's think, then, as to what might possibly give rise to a periodic effect which, at more or less fixed intervals, would place enormous strain upon life forms and weed them out with a kind of blind ruthlessness.

It has sometimes been suggested that there is a natural life-expectancy to species; that species, like individuals, have a lusty youth, a vigorous prime, a fading old age, and then a senile death. Perhaps the Great Dyings take place when the species-lifetimes of a large number of species just happen to reach the end together.

Actually, there's no evidence at all that species grow senile in the sense that individuals do, but can we put things in other terms? Instead of talking of senility and life-expectancy, let's talk of mutations.

All species are constantly subject to mutations, and mutated individuals arise in each generation. In the vast majority of cases, these mutations are for the worse and the mutated forms survive less well than do the normal. If there are enough mutations, however, and if the mutated forms are enough of a burden on the species as a whole, the species can be weakened to the point where it succumbs to its enemies. In that sense, the species may be viewed as growing "senile."

Then, too, particular species may develop a tendency for certain types of disastrous mutations. This is more likely to happen when creatures have grown so specialized that they are oversensitive to changes in the environment or in their own physiology. A creature with too elaborate a set of armor or too unbalanced a structure may pass beyond the practical with even a small change.

We ourselves are not immune. We have an extraordinarily complicated mechanism—in many stages—of blood clotting. Our blood clots with remarkable efficiency, but the complications mean that it is subject to an unusually high failure rate since there are so many stages that can go wrong. A sizable number of mutations occur in each generation of mankind that involve some imperfection in the clotting mechanism. The resultant "bleeders" cannot live long without heroic measures.

Again, the human species has developed an enormous head to house our giant brains. The female pelvis has barely kept pace and infants are born with outsize skulls that can barely squeeze through the pelvic opening and even then only at the price of distorting the still-soft cranium. In several ways, then, Homo sapiens is at the ragged edge of disaster and cannot afford a rise in mutation rate.

Suppose now there is an increase in the mutation rate. If a species or group of species is so well balanced that there are relatively few likely mutations that can result in death, it can endure that increase moderately well. If, on the other hand, a species is near disaster in some way, a sudden increase in the mutation rate might just wipe it out.

If the causes bringing about an increase in mutation rate are temporary, then only certain vulnerable species will go, while the less vulnerable ones may survive, albeit somewhat ravaged and changed.

Perhaps all the dinosaurs shared something that made them particularly vulnerable to the ravages of certain mutations. Perhaps all went (either directly or as part of the chain of life) when mutation rates climbed at the end of the Cretaceous. Those that survived (including our own ancestors) happened to be less vulnerable, that's all.

And perhaps there are additional periods of increased mutation-rates to come, and perhaps in the game of evolutionary musical chairs, we won't always be among the winners.

But what is it that happens? What would raise the mutation rate?

One answer that springs to mind is radiation. The Earth is bombarded by hard radiation of varying origin. There is the radioactivity of the crust itself, for one thing. However, there is no reason why that radioactivity should suddenly increase at particular times. In fact, the only change it can undergo, as far as we know, is that of a slow but steady decrease.

What about the radiation that bombards Earth from outer space—the Sun's radiation, and the cosmic rays from beyond the solar system?

Much of this radiation is absorbed by the atmosphere before it reaches the Earth's surface and much of it (at least the electrically charged components of it) is deflected by the Earth's magnetic field. As a result of this deflection, the Earth is surrounded by regions in which charged particles, in high density, dance back and forth along the magnetic lines of force (the Van Allen belts) and leak down into the upper atmosphere in the polar regions to form the auroras.

Clearly, if Earth's magnetic field were to vanish, charged particles (including the cosmic-ray particles) would no longer be deflected and more of them would strike the Earth's surface. The effect would be to raise the radiation level and, therefore, the mutation rate.

But could the Earth's magnetic field vanish?

Possibly! Consider the Sun, for instance. It has an 11-year sunspot cycle, as we all know. That is, the number of sunspots rises steadily, reaching a maximum, then falls to a minimum that is nearly zero, then rises to another maximum, and so on. The length of time from maximum to maximum averages 11 years, though the actual time lapse between recorded maxima has varied from 7 to 17 years.

The sunspots have magnetic fields associated with them and the orientation of the magnetic field is opposite in the two hemispheres. If in the Northern Hemisphere, spots have the north magnetic pole on top (so to speak), those in the Southern Hemisphere have the south magnetic pole on top. Then, in the next cycle, the situation switches. The Northern Hemisphere spots have the south magnetic pole on top and the Southern Hemisphere spots have the north magnetic pole there. To restore the sunspot cycle magnetically as well as numerically one must wait 22 years.

It is not certain whether this means that the Sun's general magnetic field regularly reverses polarity so that every 22 years the Sun's north magnetic pole becomes its south magnetic pole and vice versa. If this happens, one must not suppose that the magnetic axis suddenly topples and turns over. What probably happens is that the entire magnetic field weakens and declines to zero and then begins to strengthen again in the opposite direction, with sunspot minima probably coming at times of zero field. Why this happens (assuming it does happen) no one knows.

Can the same thing happen to Earth's much smaller magnetic field? Well, there are indications in the rocks (as, for instance, in the orientation of magnetized minerals) that there have been periods in Earth's history when the south magnetic pole was where the north magnetic pole now is and vice versa. Presumably, this happens because the Earth's magnetic field gradually fades to zero, then strengthens in the opposite direction.

As a matter of fact, it seems that the Earth's magnetic field has indeed been weakening during the centuries it has been under observation. The American geophysicists Keith McDonald and Robert Gunst point out it has lost 15 per cent of its strength since 1670 and at the present rate of decrease it will reach zero by 4000 A.D. Between 3500 and

4500, the magnetic field will not be strong enough to ward off any charged radiation to speak of.

We ourselves won't live to see it, of course, but a matter of two thousand years isn't long even in terms of human civilization, let alone in terms of geologic eras, and it is not something we can dismiss with a shrug.

And it is a shame, for it seems rather bad luck for us to be so close to a reversal. The last reversal, as nearly as we can tell from the rocks, may have taken place as much as 700,000 years ago.

What will happen when the magnetic field fades? Perhaps, by 3500, we will have the technological capacity to shield the Earth artificially, but suppose we don't. Will the mutation rate go up in the thousand years of non-shielding and kill the less stable or "senile" species? Will we be among the senile? Is Judgment Day coming?

Perhaps not. After all, 700,000 years ago, when the magnetic field may have reversed itself, there was no Great Dying among man's hominid ancestors. They may even have improved if the mutation rate had gone up. At least man's brain grew in volume with explosive speed (in terms of ordinary evolutionary rates of change) and one might speculate that it was the result of an unusual number of lucky mutations.

Besides, I have seen calculations which showed that even if there were no magnetic field at all and no interception of charged particles, the level of radiation at the surface of the Earth would not rise sufficiently to increase the mutation rate dangerously.

Suppose we tackle it from the other end. Forget Earth's magnetic field for a while, and ask whether the radiation bombardment might increase drastically at the source. The Sun sends out X rays from its corona as a matter of course and occasionally accompanies a giant flare with a burst of soft cosmic rays. The quantity of this radiation is too small to harm life, but what if it suddenly increased in intensity considerably.

It's not likely. The Sun could scarcely undergo the changes necessary to becoming much more active as an

X-ray and cosmic-ray emitter without becoming much more active in the emission of ultraviolet and visible light as well, and the Sun doesn't do such things.

From everything we know (or think we know) about the Sun and about stars generally, and from everything we can deduce from the fossil record, an erratic Sun is not in the cards. Our good old solar heating plant is utterly reliable and hasn't changed noticeably from eon to eon.

What about cosmic rays from sources other than the Sun? These are the only significant nonsolar samples of hard radiation that we get.

Lately, K. D. Terry of the University of Kansas and W. H. Tucker of Rice University have speculated on the possible effects of stars going supernova in the neighborhood of our solar system.

They point out that a good massive type II supernova (involving the virtually total explosion of a star ten times the mass of our Sun) would give off up to 2×10^{51} ergs of energy in the form of cosmic rays alone, emitting it all over the period of a few days at most.

Let us say that this cosmic-ray energy is delivered in the space of a week. It would then be equivalent to roughly *1 trillion times the total energy delivered by the Sun in that week.*

How much of that energy would reach us? If such a supernova were 16 light-years away, the cosmic ray energies reaching us from that vast distance would still be equal to the Sun's total radiation for that week. Undoubtedly, that would fry us all properly.

However, there are very few stars of any kind that, right now, are as close to us as 16 light-years, and of those that are none are large enough to give rise to the biggest kind of supernova. The only close star that could go supernova at all would be Sirius and that would make a rather mild one.

However, we don't have to insist on a total frying. Consider supernova explosions that take place at great distances and bathe us with smaller concentrations of cosmic rays. Those smaller concentrations might still be enough to cause trouble and there is room for many more supernovas far away than close by. The volume of space goes up as the cube of the distance and the number of supernovas within 200 light-years is two thousand times as many as the number within 16 light-years.

Terry and Tucker point out that the present dose of cosmic rays reaching the top of the atmosphere is equal to 0.03 roentgen per year, which is very little, really. And yet, judging from the frequency of supernova, their random positions and sizes, they calculate that Earth could receive a concentrated dose of 200 roentgens, thanks to supernova explosions, every 10,000,000 years or so, on the average, and considerably larger doses at correspondingly longer intervals. In the 600,000,000 years since the fossil record began there is a reasonable chance that at least one 25,000 roentgen flash (!) reached us.

Perhaps then the periodic Great Dyings in the history of life reveal the explosions of large stars within a few hundred light-years of our solar system.

And perhaps the effect is worst when such a sizable explosion just happens to come when the Earth's magnetic field is in its period of reversal and the unshielded surface gets the full benefit of the cosmic-ray frying pan. After all, our magnetic field is weak now, much weaker than at its maximum. There are probably times when even moderately strong doses of cosmic rays might not make it, but now they will, and by 3500 they will do so even more readily. A supernova that in 300,000 B.C would not have affected Earth, might lay us pretty low now.

Well, then, if we can find a record in the rocks that about 70,000,000 years ago there was a magnetic field reversal, and if we can find a record in the skies that 70,000,000 years ago there was a spectacular supernova in our neighborhood, and if there were some way of showing they were simultaneous, then I would be strongly tempted to look no further for the cause of the death of the dinosaurs.

And what about our not-too-distant descendants? Must we hold our breaths and cross our fingers for them? What if during the thousand-year interval of virtually no shielding whatever, Sirius goes supernova, or a larger, but more distant, star does it?

The chances are extremely small. As far as we know, no star within several hundred light-years is sufficiently late in its evolutionary development to make a supernova explosion likely—but, then, we don't know all there is to know about what makes a supernova explode, and when.

It just barely could be. The cosmic-ray incidence may go up enough to make a Dying, Great or Little, and there is nothing to ensure the immunity of Homo sapiens if that comes to pass.

And if we die and the crocodiles and lizards survive, there may be a kind of reptilian last laugh at our expense.

16 Counting Chromosomes

Alas, I am a square. I don't use mind-expanding drugs, I have no secret urgings toward psychedelism, I don't smoke pot (the technical term for marijuana); in fact, I don't even use alcohol or smoke tobacco. I wear my hair and sideburns moderately short, have neither beard nor mustache, dress cleanly and conservatively (if, on occasion, sloppily), and speak a reasonably precise English.

I do all this as a matter of personal choice, however, and without deep, moral convictions. I have no objections whatever to the eccentricities of others, provided they leave me to mine, and provided those eccentricities do no harm to anyone but their owners.

I consequently spring to the defense of the long-hairs against my fellow squares. In my time, I have published little essays pointing out that those who object to long hair on boys "because it makes them look like girls" ought to object to the practice of shaving, for the very same reason. (See the next chapter, for instance.) Yet they don't; but usually object to beards, too, which makes nonsense of the logic with which they try to invest their prejudices.

In fact, it seems to me that the whole business of telling the sexes apart at a glance is overrated. Why does one have to, if one doesn't have a personal interest in a particular individual? I like to quote Roland Young's "The Flea":

> And here's the happy bounding flea—
> You cannot tell the he from she.
> The sexes look alike, you see;
> But she can tell, and so can he.

So it was with some chagin that I discovered that telling a boy from a girl can be important indeed and not at all simple. In the fall of 1967, a Polish woman athlete was inspected by doctors. Her unclothed body was clearly female, but more subtle tests placed that femininity in question. The chromosome count, it seems, was wrong.

And what are these chromosomes that can thus contradict the evidence of one's eyes in so vital a matter and get away with it? Aha, as you have just guessed, I'm about to tell you at length.

The chromosomes are tiny, flexible, rodlike objects within the body's cells, visible only under the microscope. It would take some five to ten thousand of them, strung end to end, to stretch across a single inch.

Even with a microscope, they are hard to see in the living cell. Like the rest of the cell, they are translucent and light passes through them easily. For a century and a half, microscopists studied cells without seeing the chromosomes.

But then, a century ago, biologists began treating cells with some of the new dyes that chemists were then starting to synthesize. Different parts of a cell contain different chemicals; some parts therefore absorb a particular dye and some do not. The cell, under such treatment, begins to display its inner structure in technicolor splendor.

In 1880, a German biologist, Walter Flemming, was using a red dye, which clung only to certain patches inside the cell nucleus. (The cell nucleus is a small body, more or less centrally located in the cell. It was early found to control the manner in which one cell divides into two cells—the key process of growth and development.) Flemming called these colored patches of nuclear material "chromatin," from a Greek word meaning "color."

Flemming was eager to learn whether the chromatin had something to do with the nuclear control of cell division. Unfortunately, chromatin could only be seen when colored by dye, and the dye killed the cells.

What he did, then, was to study thin slices of rapidly growing tissue, in which individual cells were in all stages of division. He dyed the entire slice and caught the chromatin at every stage of that process. By putting the different stages in the right order, he could work out the details of the process. (It was like taking a set of scrambled photographic stills, putting them in the right order, and then running off a moving picture film.)

It turned out that as a cell got ready to divide, the chromatin material collected itself into what looked like a tangled mass of short pieces of cooked spaghetti. These pieces were soon given the name of "chromosomes" ("colored bodies"). At the crucial moment of division, the chromosomes separate into two equal parts, one half going to one end of the cell and the rest to the other. The cell then divided through the middle, and two new "daughter cells" were produced, each containing its own supply of chromosomes. Once division was complete, the chromosomes in each new cell broke up into patches of chromatin again.

Further study showed that a particular species of creature contained the same number of chromosomes in each of its cells (with one important exceptional case which I'll get to in a while). It is not always easy to tell what this number is, since the chromosomes tangle together, and when there are many of them, it is hard to say where one leaves off and another begins. The best attempts at counting chromosomes in human cells seemed to show, at first, that there were 48 chromosomes per cell. In 1956, however, a more painstaking count showed only 46.

Chromosome counting has now become rather simple. Cells are treated with a chemical that forces the process of cell division to stop short at just the point where the chromosomes are most clearly shaped. These cells, caught in mid-division, are then treated with a weak salt solution that causes the individual chromosomes to swell, become puffy, and move apart. They can then be counted with very little trouble, and in human cells, 46 chromosomes turns out to be correct.

But this raises a question. How can all human cells have 46 chromosomes? If the chromosomes divide into two equal parts when a cell divides, shouldn't each new cell have merely 23 chromosomes? And at the next division, shouldn't the number of chromosomes be still fewer?

No! Just before cell division, the number of chromosomes within the cell doubles. For a moment, there are 92 chromosomes in the dividing cell and, after division, each new cell has exactly half of that momentarily doubled supply and is back to 46 again. This happens at each cell division so that (again with one exception) the chromosome number remains 46 no matter how many times cells divide and share out their chromosome content.

The involvement of chromosomes in cell division is, as a matter of fact, very precise. The chromosomes aren't put together higgledy-piggledy at all. Chromatin comes together to form chromosomes of specific size and shape, and they are always formed in pairs. The human cells contain 23 pairs of chromosomes; these have been carefully numbered in order of decreasing length, from 1 to 22, with the twenty-third pair being a special case.

At the midpoint of the process of cell division, each pair of chromosomes brings about the formation of another pair exactly like itself. Since it forms a replica of itself, the process is called "replication." After replication, two complete sets of chromosomes are present; and when the cell divides, the chromosomes separate in such a way that if a particular pair moves in one direction its replica moves in the other. Each new cell ends with a complete set of chromosomes, one pair of each of the 23, with no pair missing and no pair added.

This careful division is necessary. Each chromosome, you see, is made up of strings of genes, thousands of them in each chromosome. Each gene controls the formation of a particular enzyme molecule, which in turn controls a particular chemical reaction going on in the cell.

The chromosomes, therefore, are the "chemical supervisors" of the cell. They carry its instructions, so to speak. Everything the cell does or can do is made possible by the particular nature of its enzyme supply and this is dictated by its chromosomes. Naturally, then, it is important that every new cell in the body get an exact set of chromosome pairs so that it may possess the instructions for the performance of its tasks.

(These instructions are basically the same for all cells, but they are somehow modified so that liver cells are produced in one part, brain cells in another, skin cells in still another and so on—each with widely different functions and abilities. The manner in which the chromosome instructions are modified is still a biochemical mystery, however.)

The process of replication passes chromosomes on, with precision, from cell to cell within a body. But how are they passed on from parents to offspring? How is a new body started with appropriate chromosome instructions?

This is done by way of the sex cells. The female produces egg cells; the male produces sperm cells. Each of these is distinguished from all other cells by the fact that they contain only *half* a set of chromosomes; only one chromosome of each pair. (This is the aforementioned exception to the rule that all human cells contain the same number of chromosomes.) At some stage in the formation of the sex cells a chromosome division takes place *without* prior replication. The 23 pairs simply separate, one of each pair going to one side and the rest to the other.

The egg cell is tremendously larger than the insignificantly tiny, tadpole-shaped sperm cell. That, however, need not be wounding to the male's ever-sensitive ego.

The egg cell is large because it contains a sizable food supply in addition to its chromosomes. The sperm cell contains chromosomes only. From the instructional standpoint, the two varieties of sex cells are equal.

The sex cells produced by a particular individual are not all alike. Every chromosome pair may be like every other chromosome pair, but the two individual chromosomes of the pair are not *exactly* alike. (In other words "Aa" may be like "Aa," but "A" is not like "a.") The two chromosomes of a pair may be twins as far as size and shape are concerned, but the molecular structure of the genes they contain may be significantly different.

One particular sex cell may get chromosome "A" of the first pair or it may get chromosome "a." It may get chromosome "B" of the second pair or chromosome "b" and so on. The number of different combinations that may be formed by taking, at random, one of each of 23 different pairs can be found by starting with twenty-three 2's and multiplying them together, 2^{23}. The answer is 8,388,608.

Even this is conservative, for pairs of chromosomes can sometimes wrap themselves about each other and swap pieces in any of a thousand different ways. A sex cell may get a chromosome that is mostly "A" but slightly "a."

Then, too, it is possible that a particular gene within a chromosome may undergo a change even while it is part of a living cell.

There are so many chances of variation in the chromosome pattern received by each sex cell that it is quite possible that each sex cell produced by a single individual has a slightly different set of chromosome instructions.

A new individual is formed only when a sperm cell from the male parent combines with an egg cell of the female parent to produce a "fertilized ovum." As a result of such a union of sex cells, the fertilized ovum now has a complete set of chromosomes—23 pairs, with one of each pair from its mother and the other of each pair from its father.

The possibilities of combination of sperm cell and egg cell produces a random reshuffling and recombination of genes from two separate individuals to produce a new creature with a brand new set of chromosome instructions, not like that of either parent. With all the possibilities for variation among the sex cells produced by each parent, it seems quite certain that each one of the estimated 60 billion humans who have lived since time began was distinctly different from every other one, and that this will continue for the indefinite future. (Identical twins, triplets, etc., are exceptions for they arise from a single fertilized ovum that has, for some reason, divided into two or more separate cells that then develop independently.)

This ceaseless variation in instructions from generation to generation through sorting and recombination of chromosomes is, in fact, the probable biological basis for the value of sex. Creatures can, after all, reproduce without sex, with one parent producing offspring without help; some types of species do this. When, however, two parents combine to form a new individual, the shuffling of chromosomes that takes place introduces new variations on a scale not possible otherwise. The flexibility and versatility of a species is greatly increased and it can evolve much more quickly to meet changing conditions. Sex has therefore (I am personally glad to say) replaced nonsex altogether among all but the simplest creatures.

The moment of fertilization—of the union of sperm with egg—is significant with respect to the twenty-third pair of chromosomes. It is the only one that need not be a true pair in outward appearance. In the female it is, however, the pair being composed of two fairly long chromosomes, called "X-chromosomes." A female, having two of these, can be designated as an XX.

In the male, the twenty-third pair is *not* a true pair. One of them is a normal X-chromosome, but the other is a stunted one, only about a quarter the length of the X. The short one is a "Y-chromosome" and the male is therefore an XY. Because of the sex difference, the twenty-third pair of chromosomes are often called the "sex chromosomes."

Apparently, the differences in the enzymes produced by

an XX and an XY set a body on one or the other of two different paths, one ending in a female anatomy and physiology and the other in a male version.

The Y-Chromosome in the male is largely nonfunctional, which means that the male X-chromosome has no spare as backup and males are therefore a bit more vulnerable to certain genetic abnormalities. A defective gene in an X-chromosome in a male shows up; in a female it may be countered by a whole gene in the other X-chromosome.

Some "sex-linked characteristics," such as color-blindness and hemophilia, which appear in males but rarely in females, are very noticeable. Others are not but may account for the fact that the female life span proves to be up to seven years longer than that of the male once the dangers of childbirth are banished by modern medicine.

When a female forms egg cells, the XX pair divides and each egg cell gets one X. As far as the overall shape of the chromosome content is concerned, all egg cells are therefore alike.

When a male forms sperm cells, the XY pair divides. Half the sperm cells end up with an X and half with a Y. There are, thus, two broad varieties of sperm cells formed.

In the activity that precedes fertilization, several hundred million sperm cells are released in the neighborhood of a single egg cell. The sperm cells (about half of them X and half Y) race for the egg cell, with winner take all. If an X-sperm happens to reach the egg first and fertilizes it, then the fertilized ovum is an XX and develops into a female. If a Y-sperm makes the grade, the result is an XY; that is, a male. The chances are about equal and so it happens that boys and girls are born in roughly equal numbers.

So far we are assuming that all will go well, and yet this may not be so. The process of cell division, involving the careful replication of the immensely complex chromosomes plus their precise division between two new cells, can easily go wrong and sometimes does.

Errors can take place. Sometimes these will only involve individual genes somewhere along the line of the various chromosomes. Such "mutations" can be fatal or merely disadvantageous. There are even occasions when a mutation can be favorable.

But what if it is not a submicroscopic gene that is affected, but an entire chromosome that goes wrong? In the process of cell division, with the chromosomes being yanked roughly apart, it may happen that one of them may break in two and come back together again with one piece rear-side forward. A backward chromosome-piece has its instructions reading differently, so to speak, and is not normal.

Or what if a chromosome breaks in two and does not reunite? The pieces might travel to opposite ends of the cell. One daughter cell may get a chromosome pair plus an extra piece of a third chromosome, while the other daughter cell gets not quite all of a pair.

Such chromosome aberrations are much more serious than are changes in individual genes. Chromosome breakage can involve hundreds of genes all at once. Such a wholesale blurring and slurring of instructions is almost certain to produce cells that cannot live and go through the intricate process of growth and division.

If such chromosome breakage takes place in the cells of a human adult, it need not be serious. One cell, or even a hundred cells, do not count for much among trillions. The damaged cells drop out and are replaced by those produced through correct division. In fact, since the damaged cells drop out and only the true-formed cells show up, cell division seems to be far more accurate than it may really be.

And what if the error takes place in the production of sex cells and one appears with such a chromosome aberration? In general, such a sex cell cannot develop far. Those children who do manage to be born usually lack serious chromosome aberrations and we get the idea that the processes of egg and sperm formation are much more foolproof than they really are. Heaven knows how many botched jobs are scrapped and never come to view.

A hint of the botching arises from the fact that a few aberrations manage to make it to birth and babyhood. One birth out of some five hundred, for instance, is of an infant with "Down's syndrome," or "Mongolism." (The latter name refers to the eyes, which seem to slant in such babies in a fashion associated with East Asians.) The condition involves serious mental retardation.

The cause of the syndrome was not known until 1959, when three Frenchmen, J. Lejeune, M. Gautier, and P. Turpin, counted the chromosomes in cells from three cases and found that each one had 47 chromosomes instead of 46. It turned out that the error was in the possession of *three* chromosome-21's, a normal pair plus an additional single. This was the first disease ever pinned to a chromosome aberration.

Apparently, what happens is that every once in a while, a sex cell is formed after an imperfect division of chromosome-pair 21. The sex cell that finally appears, instead of having one chromosome-21, as it should, has two or none at all. After union with another cell with the normal single chromosome present, the fertilized ovum has either three, 21-21-21, or one, 21.

The case of the three is Down's syndrome. The case of one had, until recently, never been detected. It was suspected that the possession of one presented so serious a disadvantage for the developing egg that it never reached term. But then, at the Bethesda Naval Medical Center in 1967, a mentally-retarded three-and-a-half-year-old girl was found to have a single chromosome-21. She was the first discovered case of a living human being with a missing chromosome.

Cases involving other chromosomes seem less common but are turning up. Patients with a particular type of leukemia show a tiny extra chromosome fragment in their cells. This is called the "Philadelphia chromosome" because it was first located in that city. Broken chromosomes, in general, turn up with greater than normal frequency in the cells of people with certain other (not very common) diseases.

The sex chromosomes, too, can be involved in aberrations. An egg cell can be formed with two X-chromosomes or none. A sperm cell can be formed with both an X- and Y-chromosome, or with neither. In such cases, a fertilized ovum may be formed that is XXY, or XYY, or simply X or simply Y.

Such cases are not common, perhaps because such embryos rarely complete their development. Nevertheless, they have been detected. A person born with an XXY set in his cells has the outward appearance of an underdeveloped male. On the other hand, X and XYY individuals seem to have the outward underdeveloped characteristics of a female.

The individual who made the headlines in connection with such an abnormality was Ewa Koblukowska, a tall, muscular twenty-one-year-old Polish girl. She always thought of herself as a girl, and, although flat-chested, had the sexual organs of a girl. She was, however, an excellent athlete and the question arose as to whether she might not have some male characteristics, including larger and stronger muscles than females have generally. This would be no crime, of course, but it would then be unsportsmanlike to have her compete with normal females.

Her chromosomes were counted and the six doctors (three Russians and three Hungarians) found themselves in agreement. The announcement was that there was "one chromosome too many."

Naturally, it would be useful to devise methods that would cut down on such chromosome aberrations. Failing that, it would *certainly* be advisable to avoid conditions that increase chromosome aberrations. Biologists are well acquainted with a number of these. Energetic radiation, for instance, will do so, and will produce gene mutations, too.

It is partly for this reason that world public opinion bore down so heavily against nuclear bomb tests in the open atmosphere. The radioactive particles produced might not kill outright, but they would slightly raise the mutation rate and increase the annual production of defectives of one sort or another.

But it is not radiation only that encourages mutation.

There are certain chemicals that do so—chemicals that interfere with chromosome replication and separation. Human beings are not likely to come in contact with most of the particular chemicals that chemists work with, but a few years ago there was the case of the tranquilizer thalidomide, which produced deformed babies once it was given to pregnant women. It undoubtedly produced chromosome aberrations.

Anything else? Might there be such a substance to which people had not been exposed till recently, but which was now coming into wider use?

This thought occurred to Dr. Maimon M. Cohen of the State University of New York in Buffalo. In June, 1966, after visiting a hippie hangout out of curiosity, he found himself wondering about the bizarre behavior of some of them. Were their cellular instructions being interfered with?

His work dealt with chromosome counts so he could check. Back at his laboratory, he began work with white blood cells, which could be obtained easily and in quantity from any drop of blood. He exposed some of them to a weak solution of LSD, then studied their chromosomes. He found they showed twice as many broken and abnormal chromosomes as ordinary white cells did which had not been exposed to LSD.

What about exposure to LSD inside the body? He began to test white cells from people who admitted having used LSD. So did other experimenters, after the first reports began to reach the world of science.

There seemed rather general agreement. The white cells of LSD-users had unusually high numbers of broken and abnormal chromosomes.

Was it only in the white cells, or was it in cells generally? In particular, did chromosome aberrations take place in the sex-cells of LSD-users to a greater extent than in non-users? If so, there would be more defective births among LSD-users than among others.

It is difficult to wait for births among what is still a small segment of the population, so experimenters turned to animals. Small amounts of LSD were injected into pregnant mice and, in a number of cases, there were serious abnormalities and malformations in the mouse embryos.

LSD makes its visible influence felt on the nervous system and the brain (it is for the sake of the pleasure obtained from the mental aberrations and hallucinations it produces that it is used), so it is not surprising that its effect on mice is most pronounced on the seventh day of pregnancy. It is then that the brain and nervous system are being rapidly formed; and it is brain malformations that most frequently appear in the affected embryos.

The equivalent period in human pregnancies is the third week—which generally comes before a woman knows she is pregnant and can therefore stop using the drug, if she is a user.

This adds a new dimension to LSD use, and strengthens my own feelings against it—since it is not merely an eccentricity, but is an agent of harm to individuals other than the user. Quite apart from the psychotic episodes it produces (up to murder and/or suicide) and from the danger of producing a permanent psychosis, it may increase the rate of defective births and multiply the load of human tragedy upon our planet.

The case is not yet proven, of course, but the indications are that LSD-users are undergoing the equivalent of a private bath of radiation fallout.

If so, fun may be fun—but the price can come high for themselves and higher for their unborn children.

Note: Since the above chapter was written (in January, 1968), interest in chromosome anomalies has skyrocketed. It turns out that individuals with an XYY combination are difficult people to handle. They are tall, strong and bright and are characterized by a tendency to rage and violence. Richard Speck, who killed eight nurses in Chicago in 1966, is supposed to be an XYY. A murderer was acquitted in Australia in October, 1968, because he was an XYY and therefore irresponsible. Nearly 4 per cent of the male inmates in a certain Scottish prison have turned out to be XYY. There are some estimates that XYY combinations may occur in as many as 1 man in every 3000.

I think it is only reasonable to begin thinking of a routine chromosome analysis of *everyone* and of *every* newborn child.

17 Uncertain, Coy, and Hard to Please

What with one thing and another, I have been doing a good deal of reading of Shakespeare lately* and I've noticed a great many things, including the following: Shakespeare's romantic heroines are usually much superior to his heroes in intelligence, character, and moral strength.

Juliet takes strenuous and dangerous action where Romeo merely throws himself on the ground and weeps (*Romeo and Juliet*); Portia plays a difficult and active role where Bassanio can only stand on the sidelines and wring his hands (*Merchant of Venice*); Benedick is a quick-witted fellow but he isn't a match for Beatrice (*Much Ado About Nothing*). Nor is Biron a match for Rosaline (*Love's Labour's Lost*) or Orlando a match for Rosalind (*As You Like It*). In some cases, it isn't even close. Julia is infinitely superior in every way to Proteus (*Two Gentlemen of Verona*) and Helena to Bertram (*All's Well That Ends Well*).

* Because I'm writing a book on the subject, that's why.

The only play in which Shakespeare seems to fall prey to male chauvinism is *The Taming of the Shrew* and a good case can be made out for something more subtle than merely a strong man beating down a strong woman—but I won't bother you about that here.

Yet, despite all this, I never hear of anyone objecting to Shakespeare on the ground that he presents women inaccurately. I have never heard anyone say, "Shakespeare is all right but he doesn't understand women." On the contrary, I hear nothing but praise for his heroines.

How is it, then, that Shakespeare—who, by common consent, has caught the human race at its truest and most naked under the probing and impersonal light of his genius—tells us women are, if anything, the superior of men in all that counts, and yet so many of us nevertheless remain certain that women are inferior to men. I say "us" without qualification because women, by and large, accept their own inferiority.

You may wonder why this matter concerns me. Well, it concerns me (to put it most simply) because everything concerns me. It concerns me as a science-fiction writer, especially, because science fiction involves future societies, and these, I hope, will be more rational in their treatment of 51 per cent of the human race than our present society is.

It is my belief that future societies *will* be more rational in this respect, and I want to explain my reasons for this belief. I would like to speculate about Woman in the future, in the light of what has happened to Woman in the past and what is happening to Woman in the present.

To begin with, let's admit there are certain ineradicable physiological differences between men and women. (First one to yell "Vive la différence!" leaves the room.)

But are there any differences that are primarily nonphysiological? Are there intellectual, temperamental, emotional differences that you are *sure* of and that will serve to distinguish women from men in a broad, general way? I mean differences that will hold for all cultures, as the physiological differences do, and differences that are not the result of early training.

For instance, I am not impressed by the "Women are more refined" bit, since we all know that mothers begin very early in the game to slap little hands and say, "No no, no, nice little girls don't do that."

I, myself, take the rigid position that we can never be sure about cultural influences and that the only safe distinctions we can make between the sexes are the physiological ones. Of these, I recognize two:

1. Most men are physically larger and physically stronger than most women.

2. Women get pregnant, bear babies, and suckle them. Men don't.

What can we deduce from these two differences *alone?* It seems to me that this is enough to put women at a clear disadvantage with respect to men in a primitive hunting society, which is all there was prior to, say, 10,000 B.C.

Women, after all, would be not quite as capable at the rougher aspects of hunting and would be further handicapped by a certain ungainliness during pregnancy and certain distractions while taking care of infants. In a catch-as-catch-can jostle for food, she would come up at the rear every time.

It would be convenient for a woman to have some man see to it that she was thrown a haunch after the hunting was over and then see to it, further, that some other man didn't take it away from her. A primitive hunter would scarcely do this out of humanitarian philosophy; he would have to be bribed into it. I suppose you're all ahead of me in guessing that the obvious bribe is sex.

I visualize a Stone Age treaty of mutual assistance between Man and Woman—sex for food—and as a result of this kind of togetherness, children are reared and the generations continue.*

* After this article appeared, an anthropologist named Charlotte O. Kursh wrote me a long and fascinating letter that made it quite clear that I had dreadfully oversimplified the situation described here, that hunting was not the only food-

I don't see that any of the nobler passions can possibly have had anything to do with this. I doubt that anything we would recognize as "love" was present in the Stone Age, for romantic love seems to have been a rather late invention and to be anything but widespread even today. (I once read that the Hollywood notion of romantic love was invented by the medieval Arabs and was spread to our own Western society by the Provençal troubadours.)

As for the concern of a father for his children, forget it. There seem definite indications that men did not really understand the connection between sexual intercourse and children until nearly historic times. Mother love may have its basis in physiology (the pleasure of suckling, for instance) but I strongly suspect that father love, however real it may be, is cultural in origin.

Although the arrangement of sex for food seems a pretty reasonable *quid pro quo*, it isn't. It is a terribly unfair arrangement because one side can break the agreement with impunity and the other cannot. If a woman punishes by withholding sex and a man by withholding food, which side will win out? *Lysistrata* to the contrary, a week without sex is a lot easier than a week without food. Furthermore, a man who tires of this mutual strike can take what he wants by force; a woman can't.

It seems to me, then, that for definite physiological reasons, the original association of men and women was a strictly unequal one, with man in the role of master and woman in the role of slave.

This is not to say that a clever woman, even in Stone Age times, might not have managed to wheedle and cajole a man into letting her have her own way. And we all know that this is certainly true nowadays, but wheedling and cajolery are slave weapons. If you, Proud Reader, are

source, and that questions of status were even more important than sex. Once one substituted "status-for-food" for "sex-for-food" she found she tended to agree with what followed. So, with this warning to take my anthropology with a grain of salt, let's continue.

a man and don't see this, I would suggest you try to wheedle and cajole your boss into giving you a raise, or wheedle and cajole a friend into letting you have your way, and see what happens to your self-respect.

In any master-slave relationship the master does only that portion of the work that he likes to do or that the slave cannot do; all else is reserved for the slave. It is indeed frozen into the slaves' duties not only by custom but by stern social law which defines slaves' work as unfit for free men to do.

Suppose we divide work into "big-muscle" and "little-muscle." Men would do the "big-muscle" work because he would have to and the women would then do the "little-muscle" work. Let's face it; this is usually (not always) a good deal for men because there is far more "little-muscle" work to do. ("Men work from sun to sun; women's work is never done," the old saying goes.)

Sometimes, in fact, there is no "big-muscle" work to do at all. In that case the Indian brave sits around and watches the squaw work—a situation that is true for many non-Indian braves who sit and watch their non-Indian squaws work.* Their excuse is, of course, that as proud and gorgeous males they can scarcely be expected to do "women's work."

The social apparatus of man-master and woman-slave was carried right into the most admired cultures of antiquity and was never questioned there. To the Athenians of the Golden Age, women were inferior creatures, only dubiously superior to domestic animals, and with nothing in the way of human rights. To the cultivated Athenian, it seemed virtually self-evident that male homosexuality was the highest form of love, since that was the only way in which a human being (male, that is) could love an equal. Of course, if he wanted children, he had to turn to a woman, but so what; if he wanted transportation, he turned to his horse.

As for that other great culture of the past, the Hebrew,

---

* Of course, if they are too chivalrous to watch a woman do all the work, they can always close their eyes. That will even give them a chance to sleep.

it is quite obvious that the Bible accepts male superiority as a matter of course. It is not even a subject for discussion at any point.

In fact, by introducing the story of Adam and Eve, it has done more for woman's misery than any other book in history. The tale has enabled dozens of generations of men to blame everything on women. It has made it possible for a great many holy men of the past to speak of women in terms that a miserable sinner like myself would hesitate to use in referring to mad dogs.

In the ten commandments themselves, women are casually lumped with other forms of property, animate and inanimate. It says, in Exodus 20:17: "Thou shalt not covet thy neighbour's house, thou shalt not covet thy neighbour's wife, nor his manservant, nor his maidservant, nor his ox, nor his ass, nor anything that is thy neighbour's."

Nor is the New Testament any better. There are a number of quotations I can choose from, but I will give you this one from Ephesians 5:22-24: "Wives, submit yourselves unto your own husbands, as unto the Lord. For the husband is the head of the wife, even as Christ is the head of the church: and he is the saviour of the body. Therefore as the church is subject unto Christ, so let the wives be to their own husbands in every thing."

This seems to me to aspire to a change in the social arrangement of man/woman from master/slave to God-/creature.

I don't deny that there are many passages in both the Old and New Testaments that praise and dignify womankind. (For example, there is the Book of Ruth.) The trouble is, though, that in the social history of our species, those passages of the Bible which taught feminine wickedness and inferiority were by far the more influential. To the self-interest that led men to tighten the chains about women was added the most formidable of religious injunctions.

The situation has not utterly changed in its essence, even now. Women have attained a certain equality before the law—but only in our own century, even here in the United States. Think how shameful it is that no woman, however intelligent and educated, could vote in a national election until 1920—despite the fact that the vote was freely granted to every drunkard and moron, provided only that he happened to be male.

Yet even so—though women can vote, and hold property, and even own their own bodies—all the social apparatus of inferiority remains.

Any man can tell you that a woman is intuitive rather than logical, emotional rather than reasonable, finicky rather than creative, refined rather than vigorous. They don't understand politics, can't add a column of figures, drive cars poorly, shriek with terror at mice, and so on and so on and so on.

Because women are all these things how can they be allowed an equal share with men in the important tasks of running industry, government, society?

Such an attitude is self-fulfilling, too.

We begin by teaching a young man that he is superior to young women, and this is comforting for him. He is automatically in the top half of the human race, whatever his shortcomings may be. Anything that tends to disturb this notion threatens not only his personal self-respect but his virility.

This means that if a woman happens to be more intelligent than a particular man in whom she is (for some arcane reason) interested, she must never, for her very life, reveal the fact. No sexual attraction can then overcome the mortal injury he receives in the very seat and core of his masculine pride, and she loses him.

On the other hand, there is something infinitely relieving to a man in the sight of a woman who is, manifestly, inferior to himself. It is for that reason that a silly woman seems "cute." The more pronouncedly male-chauvinistic a society the more highly valued is silliness in a woman.

Through long centuries, women have had to interest men somehow, if they were to achieve any economic security and social status at all, and so those who were not stupid and silly by nature had to carefully cultivate such stupidity and silliness until it came natural and they forgot they ever were intelligent.

It is my feeling that all the emotional and temperamental distinctions between men and women are of cultural origin, and that they serve the important function of maintaining the man/woman master/slave arrangement.

It seems to me that any clear look at social history shows this—and shows, moreover, that the feminine "temperament" jumps through hoops whenever that is necessary to suit man's convenience.

What was ever more feminine than Victorian womanhood, with its delicacy and modesty, its blushes and catchings of breath, its incredible refinement and its constant need for the smelling salts to overcome a deplorable tendency to faint? Was there ever a sillier toy than the stereotype of the Victorian woman; was there ever a greater insult to the dignity of Homo sapiens?

But you can see why the Victorian woman (or a rough approximation of her) had to exist in the late nineteenth century. It was a time when among the upper classes, there was no "little-muscle" work for her to do since servants did it. The alternative was to let her use her spare time in joining men in their work, or to have her do nothing. Firmly, men had her do nothing (except for such make-work nothings as embroidery and hack piano-playing). Women were even encouraged to wear clothes that hampered their physical movements to the point where they could scarcely walk or breathe.

What was left to them, then, but a kind of ferocious boredom that brought out the worst aspects of the human temperament, and made them so unfit an object even for sex, that they were carefully taught that sex was dirty and evil so that their husbands could go elsewhere for their pleasures.

But in this very same era, no one ever thought of applying the same toy-dog characteristics to the women of the lower classes. There was plenty of "little-muscle" work for them to do and since they had no time for fainting and refinement, the feminine temperament made the necessary adjustment and they did without either fainting or refinement.

The pioneer women of the American West not only cleaned house, cooked, and bore baby after baby, but they grabbed up rifles to fight off Indians when necessary. I strongly suspect they were also hitched to the plow on such occasions as the horse needed a rest, or the tractor was being polished. And this was in Victorian times.

We see it all about us even now. It's an article of faith that women just aren't any good at even the simplest arithmetic. You know how those cute little dears can't balance a checkbook. When I was a kid, all bank tellers were male for that very reason. But then it got hard to hire male bank tellers. Now 90 per cent of them are female and apparently they can add up figures and balance checkbooks after all.

There was a time all nurses were males because everyone knew that women were simply too delicate and refined for such work. When the economic necessities made it important to hire females as nurses, it turned out they weren't all that delicate and refined after all. (Now nursing is "woman's work" that a proud man wouldn't do.)

Doctors and engineers are almost always men—until some sort of social or economic crunch comes—and then the female temperament makes the necessary change and, as in the Soviet Union, women become doctors and engineers in great numbers.

What it amounts to is best expressed in a well-known verse by Sir Walter Scott:

> *O woman! in our hours of ease,*
> *Uncertain, coy, and hard to please,*
>
> . . .
>
> *When pain and anguish wring the brow,*
> *A ministering angel thou!*

Most women seem to think this is a very touching and wonderful tribute to them, but I think that it is a rather bald exhibition of the fact that when man is relaxing he wants a toy and when he is in trouble he wants a slave and woman is on instant call for either role.

What if pain and anguish wring *her* brow? Who's *her* ministering angel? Why, another woman who is hired for the occasion.

But let's not slip to the other extreme either. During the fight for women's votes, the male chauvinists said that this would wreck the nation since women had no feeling for politics and would merely be manipulated by their menfolks (or by their priests, or by any political quack with a scalpful of curls and a mouthful of teeth).

Feminists, on the other hand, said that when women brought their gentleness and refinement and honesty to the polling booth, all graft, corruption, and war would be brought to an end.

You know what happened when women got the vote? *Nothing.* It turned out that women were no stupider than men—and no wiser, either.

What of the future? Will women gain true equality?

Not if basic conditions continue as they have ever since Homo sapiens became a species. Men won't voluntarily give up their advantage. Masters never do. Sometimes they are forced to do so by violent revolution of one sort or another. Sometimes they are forced to do so by their wise foresight of a coming violent revolution.

An *individual* may give up an advantage out of a mere sense of decency, but such are always in the minority and a group as a whole never does.

Indeed, in the present case, the strongest proponents of the status quo are the women themselves (at least most of them). They have played the role so long they would feel chills about the wrists and ankles if the chains were struck off. And they have grown so used to the petty rewards (the tipped hat, the offered elbow, the smirk and leer, and, most of all, the freedom to be silly) that they won't exchange them for freedom. Who is hardest on the independent-minded woman who defies the slave-conventions? Other women, of course, playing the fink on behalf of men.

Yet things will change even so, because the basic conditions that underlie woman's historic position are changing.

What was the first essential difference between men and women?

1. Most men are physically larger and physically stronger than most women.

So? What of that today. Rape is a crime and so is physical mayhem even when only directed against women. That doesn't stop such practices altogether, but it does keep them from being the universal masculine game they once were.

And does it matter that men are larger and stronger, in the economic sense? Is a woman too small and weak to earn a living? Does she have to crawl into the protecting neck-clutch of a male, however stupid or distasteful he may be, for the equivalent of the haunch of the kill?

Nonsense! "Big-muscle" jobs are steadily disappearing and only "little-muscle" jobs are left. We don't dig ditches any more, we push buttons and let machines dig ditches. The world is being computerized and there is nothing a man can do in the way of pushing paper, sorting cards, and twiddling contacts, that a woman can't do just as well.

In fact, littleness may be at a premium. Smaller and slenderer fingers may be just what is wanted.

More and more, women will learn they need only offer sex for sex and love for love, and nevermore sex for food. I can think of nothing that will dignify sex more than this change, or more quickly do away with the degrading master/slave existence of "the double standard."

But how about the second difference:.

2. Women get pregnant, bear babies, and suckle them. Men don't.

I frequently hear tell that women have a "nest-building" instinct, that they really *want* to take care of a man and immolate themselves for his sake. Maybe so, under conditions as they used to be. But how about now?

With the population explosion becoming more and more of a cliff-hanger for all mankind, we will, before the end of the century, have evolved a new attitude toward babies or our culture will die.

It will become perfectly all right for a woman not to have babies. The stifling social pressure to become a "wife and mother" will lift and that will mean even more than the lifting of the economic pressure. Thanks to the pill, the burden of babies can be lifted without the abandonment of sex.

This doesn't mean women *won't* have babies; it means merely they won't *have* to have babies.

In fact, I feel that female slavery and the population explosion go hand in hand. Keep a woman in subjection and the only way a man will feel safe is to keep her "barefoot and pregnant." If she has nothing to do except undignified and repetitive labor, a woman will want baby after baby as the only escape to something else.

On the other hand, make women truly free and the population explosion will stop of its own accord. Few women would want to sacrifice their freedom for the sake of numerous babies. And don't say "No" too quickly; feminine freedom has never been truly tried, but it must be significant that the birth rate is highest where the social position of women is lowest.

In the twenty-first century, then, I predict that women will be completely free for the first time in the history of the species.

Nor am I afraid of the counter-prediction that all things go in cycles and that the clearly visible trend toward feminine emancipation will give way to a swing back to a kind of neo-Victorianism.

Effects can be cyclic, yes—but only if causes are cyclic, and the basic causes here are non-cyclic, barring worldwide thermonuclear war.

In order for the pendulum to swing back toward feminine slavery, there would have to be an increase in "big-muscle jobs" that only men could do. Women must begin once more to fear starvation without a man to work for them. Well, do you think the present trend toward computerization and social security will reverse itself short of global catastrophe? Honestly?

In order for the pendulum to swing back, there would have to be a continuation of the desire for large families and lots of children. There's no other way of keeping women contented with her slavery on a large scale (or too busy to think about it, which amounts to the same thing). Given our present population explosion and the situation as it will be by 2000, do you honestly expect women to be put to work breeding baby after baby?

So the trend toward woman's freedom is irreversible.

There's the beginning of it right now and it is well established. Do you think that the present era of increasing sexual permissiveness (almost everywhere in the world) is just a temporary breakdown in our moral fiber and that a little government action will restore the stern virtues of our ancestors?

Don't you believe it. Sex has been divorced from babies, and it will continue to be so, since sex can't possibly be suppressed and babies can't possibly be encouraged. Vote for whom you please but the "sexual revolution" will continue.

Or take even something so apparently trivial as the new fad of hairiness in man. (I've just grown a pair of absolutely magnificent sideburns myself.) Sure, it will change in details, but what it really stands for is the breakdown of trivial distinctions between the sexes.

It is indeed this which disturbs the conventional. Over and over, I hear them complain that some particular long-haired boy looks just like a girl. And then they say, "You can't tell them apart any more!"

This always makes me wonder why it is so important to tell a boy from a girl at a glance, unless one has some personal object in view where the sex makes a difference. You can't tell at a glance whether a particular person is Catholic, Protestant, or Jew; whether he/she is a piano player or a poker player, an engineer or an artist, intelligent or stupid.

After all, if it were *really* important to tell the sexes

apart at the distance of several blocks with one quick glance, why not make use of Nature's distinction? That is *not* long hair since both sexes in all cultures grow hair of approximately equal length. On the other hand, men always have more facial hair than women; the difference is sometimes extreme. (My wife, poor thing, couldn't grow sideburns even if she tried.)

Well, then, should all men grow beards? Yet the very same conventional people who object to long hair on a man, also object to beards. *Any* change unsettles them, so when change becomes necessary, conventional people must be ignored.

But *why* this fetish of short hair for men and long hair for women, or, for that matter, pants for men and skirts for women, shirts for men and blouses for women? Why a set of artificial distinctions to exaggerate the natural ones? Why the sense of disturbance when the distinctions are blurred?

Can it be that the loud and gaudy distinction of dress and hair between the two sexes is another sign of the master-slave relationship? No master wants to be mistaken for a slave at any distance, or have a slave mistaken for a master, either. In slave societies, slaves are always carefully distinguished (by a pigtail when the Manchus ruled China, by a yellow Star of David when the Nazis ruled Germany, and so on). We ourselves tend to forget this since our most conspicuous non-female slaves had a distinctive skin color and required very little else to mark them.

In the society of sexual equality that is coming, then, there will be a blurring of artificial distinctions between the sexes, a blurring that is already on the way. But so what? A particular boy will know who his particular girl is and vice versa, and if someone else is not part of the relationship what does he/she care which is which?

I say we can't beat the trend and we should therefore join it. I say it may even be the most wonderful thing that has ever happened to mankind.

I think the Greeks *were* right in a way and that it *is* much better to love an equal. And if that be so, why not hasten the time when we heterosexuals can have love at its best?

DISCUS BOOKS
DISTINGUISHED NON-FICTION

A SELECTION OF RECENT TITLES

DRT 7-76

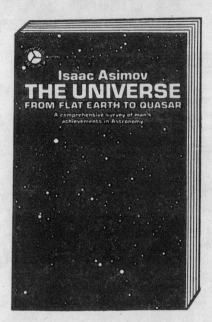